Developments in Primatology: Progress and Prospects

Series Editor: Louise Barrett

This book series melds the facts of organic diversity with the continuity of the evolutionary process. The volumes in this series will exemplify the diversity of theoretical perspectives and methodological approaches currently employed by primatologists and physical anthropologists. Specific coverage includes primate behavior in natural habitats and captive settings; primate ecology and conservation; functional morphology and developmental biology of primates; primate systematics; genetic and phenotypic differences among living primates and paleoprimatology.

More information about this series at http://www.springer.com/series/5852

Claire A. Kirchhoff

Life and Death in the Gombe Chimpanzees

Skeletal Analysis as an Insight into Life History

Springer

Claire A. Kirchhoff
Department of Biomedical Sciences
Marquette University
Milwaukee, WI, USA

ISSN 1574-3489 ISSN 1574-3497 (electronic)
Developments in Primatology: Progress and Prospects
ISBN 978-3-030-18354-7 ISBN 978-3-030-18355-4 (eBook)
https://doi.org/10.1007/978-3-030-18355-4

This Springer imprint is published by the registered company Springer Nature Switzerland AG.
The registered company address is: Gewerbestrasse 11, 6330 Cham, Switzerland

For my grandparents

Preface

I began working with the Gombe chimpanzee skeletal collection in 2008. Ten years on, it is easy for me to see, with the benefit of hindsight, the ways in which data collection and analysis could be improved. While the shoulders of giants and the dedication of many have made the presentation of this work possible in a format I think uniquely suited to provide a narrative for these skeletons, any shortcomings or errors are mine.

It is humbling to be a small part of the story that continues to unfold from Gombe, and I therefore present my findings humbly, keenly aware of how small a piece of that story this is. The chimpanzees of Gombe have a lot to tell us—about east African ecology, about primates, about skeletal analysis, and about ourselves. My greatest hope for this book is that it might serve as a reminder of the value of integrating hard tissue data with other lines of evidence and inspire others to preserve and study the skeletons of wild primates with known life histories.

Milwaukee, WI, USA — Claire A. Kirchhoff

Acknowledgments

I am deeply indebted to many individuals and organizations who have made this project possible.

First, thank you to Dr. Michael L. Wilson, who connected me with the Gombe research team, guided me through acquiring research permissions, and handed me Goblin's precious skeleton on my first day at the Park, as though he were a treasured friend. This book results from his continued mentoring even after I graduated.

I am also grateful to my entire dissertation advisory committee, including Dr. Kieran McNulty, Dr. Galin Jones, and my advisor, Dr. Martha Tappen, who has supported my career and encouraged me to be a scientist since I was an undergraduate at the University of Minnesota.

What makes the Gombe chimpanzee skeletal collection compelling is the opportunity to integrate multiple lines of evidence. Toward this end, this project would not have been possible without the generosity of Dr. Anne Pusey in providing me with dominance rank data. I am also indebted to Dr. Joseph T. Feldblum. Dr. Karen Terio, Dr. Elizabeth Lonsdorf, and Dr. Jane Raphael also generously shared their insights on soft tissue lesions and demographic data. Ongoing behavioral research is coordinated by Dr. Deus C. Mjungu, the director of chimpanzee research at Gombe. His dedication and the dedication of the field observers affiliated with Gombe National Park make possible a detailed and rich behavioral record of the chimpanzees who live there. Among them are Yahaya Almasi, Caroly Alberto, Bashiru Butoki, Lamba Hilali, Baraka Gilagiza, Matendo Msafiri, Gabo Paulo, Gadison Titya, Methodi Vyampi, and Simon Yohana. Asante sana!

It is thanks to the efforts of many people associated with Gombe National Park over the course of several decades that this skeletal collection exists. I owe much gratitude to Dr. Anthony Collins and Dr. Shadrack Kamenya for access to the skeletons currently curated by Gombe Stream Research Center. Dr. Collins has been a tireless encourager since I first visited Gombe. I also thank Dr. Anna Mosser, who served as the chimpanzee research director at Gombe while I had the privilege of visiting the Park. In addition, for many years, Dr. Adrienne Zihlman studied and curated the skeletons now housed at the University of Minnesota. I also thank the University of Minnesota Evolutionary Anthropology Labs for allowing me access to the skeletal collection, even after completing my degree, to collect some additional data. The lab director, Matthew Edling, coordinated these visits and made me feel welcome.

Various agencies granted me permission to undertake this research, including the Jane Goodall Institute and Dr. Goodall, who remains a light of inspiration to our world. In addition, I was graciously granted permits by the Tanzanian Research Institute, Tanzania National Parks, and the Tanzanian Commission for Science and Technology. Funding for this project was made possible thanks to the University of Minnesota Graduate Research Partnership Program, the University of Minnesota Thesis Research Grant, and lots of support and funding over the course of 7 years from the University of Minnesota, Department of Anthropology.

Dr. Scott Legge and Rebecca Nockerts provided valuable feedback on portions of this manuscript at key points during its development.

Finally, I would like to acknowledge my family: the unconditional love and support from my parents, Kathleen and Donn Kirchhoff, and the inspiring examples set by my grandparents fueled my education. It is also fair to say that my dissertation might not have been submitted with proper formatting if not for the intervention of my sister, Carolyn Kirchhoff Reeves, MBA. Last, I owe so much to my partner, Dr. Matthew D. Hunstiger, for making my career possible, for always standing beside me, and for his many sacrifices in the interest of my happiness.

Contents

Chapter 1
The Gombe Skeletal Sample and Case Studies

1.1 Introduction

Evolutionary mechanisms such as natural selection act at the level of the individual. The study of individual life histories is therefore of particular interest when considering the selective pressures that act upon an organism. The Gombe chimpanzees provide a unique opportunity to explore these mechanisms, both through the vast amounts of behavioral data collected over the course of 50 years and how these life histories impact the hard tissues. The study of the relationship between behavior and the hard tissues is of special relevance for a complete picture of skeletal biology, because it is only through the study of skeletons with known life histories that it is possible to interpret unknown skeletal material with any accuracy.

Since 1960, Dr. Jane Goodall and colleagues have studied the chimpanzees of Gombe National Park, Tanzania. The skeletal materials obtained from chimpanzees who have died during the course of this ongoing study represent an unparalleled opportunity to better understand the effects of a chimpanzees' life history and behavior on their skeletons. An extraordinary amount of detailed information is available about the lives of Gombe chimpanzees (e.g., Goodall 1986) whose skeletons have been preserved, and it is rare for life history data to be available for the same animals for whom skeletal data are available. Because of the nature of their acquisition, museum skeletal collections do not usually include life history data (Harman 2005) and laboratory chimpanzees, for whom behavioral data are often available, do not allow for appropriate comparisons, or make good proxies for the study of primate evolution. Reasons for the decreased value of captive ape skeletons for research of this nature include the very different stresses associated with captivity compared to the wild (e.g., less nutritional stress, access to veterinary care, and often more psycho-social stress), and that captive chimpanzees achieve larger ultimate body sizes and grow up faster than their wild counterparts (Zihlman et al. 2004a). Reliable sets

C. A. Kirchhoff, *Life and Death in the Gombe Chimpanzees*,
Developments in Primatology: Progress and Prospects,
https://doi.org/10.1007/978-3-030-18355-4_1

of both skeletal and behavioral data are only obtained for wild animals from behavioral research stations such as Gombe National Park in Tanzania.

We are indeed fortunate to be able to study the complete lives—from birth to death and beyond—of chimpanzees from Gombe. Former alpha-male Goblin is a prime example of this phenomenon. First observed as an infant just a few days old, the entire course of his life has been carefully recorded and analyzed. I am pleased to present a description of Goblin's skeleton here for the first time, one of 49 chimpanzees represented in the skeletal collection from Gombe National Park.

In addition to providing a more complete life history for a particular organism, individual-level analysis can also make species-level trends apparent, especially over relatively long periods of time. This makes the Gombe skeletal collection especially relevant, as it includes chimpanzees of a wide variety of ages with diverse causes of death who have died over the course of half a century. Because chimpanzees are so long lived, it is only after considerable time has passed that it is possible to more fully understand their life histories, the selective pressures that act upon individual chimpanzees, and the common sources of morbidity and mortality of the species. Skeletal evidence gives us relevant information about which sources of morbidity and mortality are reflected by the hard tissues. This is relevant for interpreting the fossil record, determining which illnesses/injuries are likely to appear on the skeleton, and helping to identify individual causes of death. In addition, continued re-examination of a growing skeletal collection can aid in the revelation of long-term trends that would not otherwise be apparent. For example, the larger sample presented here allows for statistical analyses not possible with earlier, smaller sample sizes.

Several of the animals described have been studied previously, and have been examined in a wide range of useful research on morphological and size variation, trauma, and pathology. This study's primary purpose is not to negate previous studies, but rather to engage the sample in a quantitative way that was not previously possible due to a smaller sample size, and to compare the Gombe chimpanzees with more recently published studies.

1.2 Previous Studies of the Gombe Skeletal Collection

Previous studies of the Gombe chimpanzee skeletal material have focused primarily on chimpanzees who died between the years 1966 and 1987 and have already yielded a great deal of valuable information. Previous research covers a wide variety of topics, including studies of degenerative joint disease (Jurmain 1989, 2000), trauma (Jurmain 1997, 1989), generalized skeletal pathology (Jurmain 1989), dental pathology (Kilgore 1989), body size (Morbeck and Zihlman 1989), skeletal age changes (Sumner et al. 1989; Morbeck et al. 2002), the skeletal effects of poliomyelitis (Morbeck et al. 1991), sex differences in the pelvis (Morbeck et al. 1992) and the vertebrae (Galloway et al. 1996), asymmetry due to hand preference (Morbeck et al. 1994), dental maturation (Zihlman et al. 2004a, b), and individual life histories as reflected in the skeleton (Zihlman et al. 1990). The sample presented here includes animals who died as recently at 2008, is significantly larger than the earlier sample, and therefore allows for statistical analyses that were not previously possible. The sample also includes chimpanzees of a wider variety of ages, which allows for a closer look at behavior's relationship to the skeleton at multiple stages of life history. I examine the relationship between skeletal (including dental) lesions and variations versus the life history variables age, sex, and dominance rank.

1.3 Dominance Rank

Dominance rank plays an important role in the social lives of chimpanzees. It affects access to resources (Pusey et al. 2005; Murray et al. 2006), inter-individual relationships (Foster et al. 2009; Goodall 1986), and reproductive success (Pusey et al. 1997; Constable et al. 2001; Wroblewski et al. 2009), among other variables. It seems obvious to assume that higher ranking chimpanzees are "healthier" in some way than their lower-ranking counterparts, but long-term observations and analysis of chimpanzee behavior reveal that the relationship between rank and various markers of health is complex.

For example, body mass in living chimpanzees at Gombe is affected by food availability. Chimpanzees commonly experience a seasonal fluctuation in body mass that is correlated with the availability of ripe fruit. Higher-ranking females chimpanzees tend to have larger body masses than lower-ranking females. This is not true for male chimpanzees, however, for whom dominance rank and body mass are not correlated (Pusey et al. 2005).

Hormone analysis, particularly of cortisol, is a long-standing technique for evaluating stress levels in wild primates as well as how stress levels are affected by an animal's own social status and the statuses of conspecifics (e.g., Sapolsky 2005). If higher-ranking primates are healthier, we might expect that they would experience lower levels of stress. Conversely, if it is stressful to be high ranking, because of increased agonistic encounters, for example, we might expect that lower-ranking chimpanzees would have lower cortisol levels, reflecting lower levels of stress.

The interaction between cortisol levels and social status or dominance rank is complex; it is affected by a large number of variables, including the social structure of a particular species (Sapolsky 2005). In other words, it is not safe to assume that all species will demonstrate the same relationship between rank and cortisol levels. This relationship may also change over time, even within the same group of animals, because of factors such as divergent dominance "styles" of different animals (Foster et al. 2009; Ray and Sapolsky 1992) or instability in the dominance hierarchy (Sapolsky 1983; Bergman et al. 2005). Dominance rank often changes over the course of a chimpanzee's life, particularly for males (Kawanaka 1990; Takahata 1990), meaning that stress levels will not be consistent throughout life history. In addition, dominance rank is not the only variable that influences cortisol levels (Lane 2006), so the effects of diet, time of day, estrus cycle (Ibid.), energetic costs

(Muller and Wrangham 2004), and other behaviors (e.g., rate of consortships and number of aggressive interactions) (Virgin et al. 1997; Muller and Wrangham 2004; Schino et al. 2007), and other behavioral styles or personality traits (Anestis 2005; Anestis et al. 2006) should be considered in addition to the effects of the social hierarchy.

In chimpanzees in particular, high male dominance rank has been found to positively correlate with higher urinary cortisol levels, indicating that maintaining high rank is costly in some way. Rates of male aggression as well as lower abundance of fruit were both correlated with higher cortisol levels in male chimpanzees (Muller and Wrangham 2004), suggesting that both social and energetic factors influence stress. In addition, lower-ranking male chimpanzees have higher urinary C-peptide concentrations, indicating a better energetic condition than higher-ranking males. The benefits of access to prime feeding sites that high rank confers therefore seem to be more than offset by the energetic costs of the stress of the aggressive interactions needed to achieve and maintain high rank (Thompson et al. 2009). In other words, dominant male chimpanzees may spend more time maintaining their rank and less time feeding (Muller and Wrangham 2004; Thompson et al. 2009), and the benefits to reproductive success resulting from this strategy are presumed to outweigh the energetic costs (Muller and Wrangham 2004). High-ranking female chimpanzees, on the other hand, do not have elevated urinary cortisol levels compared to lower-ranking females. In particular, immigrant female chimpanzees, who are low ranking during the period immediately following immigration, experience high levels of intra-sexual aggression. This is correlated with high stress levels as measured by urinary cortisol (Kahlenberg et al. 2008).

If we now have improved data on how hormone levels are affected by specific stressors as well as how these variables correlate to dominance rank, this still leaves many questions about longer-term trends in health status as well as the direction of causality. Hormone levels of any kind are, by nature, more ephemeral and time specific than skeletal markers of health and stress. Skeletal trauma and pathology incidents accumulate over the lifetime of an organism; even old injuries may still leave a sign on the skeleton. Examining skeletal markers of health and stress therefore has great potential for studying the cumulative effects of stress over the course of a chimpanzee's entire life, and how these effects may be related to dominance rank.

A correlation between something like cortisol levels and dominance rank does not contain information on causality. Does having a higher rank result in higher cortisol levels for male chimpanzees? Alternately, could elevated cortisol levels (and perhaps a greater tolerance for stress (Chichinadze and Chichinadze 2008)) make a chimpanzee more likely to achieve high rank? If skeletal markers of stress, which can accumulate during early development, correlate well with dominance rank achieved late in life, it may be possible to determine direction of causality. Are chimpanzees with higher levels of developmental stress more or less likely to achieve high rank? I examine this question in relationship to potential stress events accumulated over a relatively long period of time (skeletal trauma and pathology).

From previous research on cortisol levels as a measure of stress and its relationship to dominance rank, I expect that:

1. Higher ranking males will have more skeletal lesions than their low-ranking counterparts.
2. Lower-ranking females will have more skeletal lesions than their high-ranking counterparts.

Trauma and pathology reflect sources of morbidity such as falls, inter-individual violence, and joint disease (Lovell 1991, 1990; Jurmain 1989, 1997; Carter et al. 2008; Kilgore 1989), and also help to paint the story of a chimpanzee's life history (Zihlman et al. 1990). Studying these phenomena at the level of the individual is particularly useful because natural selection acts at this level. Understanding the relationship between fitness and instances of trauma and pathology is therefore of great interest. I expect, because of the high costs of agonistic encounters associated with high rank in male chimpanzees, that high-ranking males (or males who once achieved high rank) will have the highest levels of skeletal trauma of any demographic group. Low-ranking females should be the next most affected group.

1.4 Skeletal Analysis

In this section, I discuss the entire Gombe chimpanzee skeletal collection, skeletonization procedure, the assessment of dominance rank from behavioral data, and methodologies for recording and analyzing trauma and pathology.

Forty-nine (49) chimpanzees are represented in the Gombe skeletal collection; of these, 37 are complete or nearly complete skeletons. Most of the chimpanzees have known life histories, and many of those who were born after 1960 include complete life histories from birth until death. Protocols for behavioral observation have been outlined elsewhere (e.g., Goodall 1986).

Table 1.1 is a demographic summary of the skeletal collection. Age categories in Table 1.1 are after Goodall (1986). Table 1.2 is a summary of the chimpanzees presented here, including completeness of skeleton, birth dates, and death dates (GSRC).

Table 1.1 Demographic summary of the Gombe chimpanzee skeletal sample

Age category	Females	Males	Unknown sex	Total
Infant (0–4 years)	2	8	3	13
Juvenile (5–7 years)	0	0	0	0
Adolescent (females: 8–14 or 15, males: 8–15)	1	7	0	8
Adult (females: 14 or 15–33, males: 16–33)	10	9	0	19
Old adult (34+)	4	5	0	9
Total	17	28	3	49

Table 1.2 The Gombe chimpanzee skeletal sample

Chimpanzee	Sex	Community	Birth date	Death date	Age at death (years)	Cause of death	Completeness of remains
Andromeda	F	Mitumba	18 Nov 2004	13 Aug 2005	0.73	Intraspecific aggression	Nearly complete (perimortem trauma)
Atlas	M	Kasekela	3 Jul 1959[a]	14 Jan 1999	31.3[a]	Illness	Complete
Beethoven?	M	Kasekela	4 Jul 1959[a]	15 Dec 2002	33.45[a]	Illness	Cranium only
Bwavi male	M	Kalande	n/a	1994 or 1995	40+?[a]	Illness	Complete
Charlie	M	Kahama	2 Jul 1951[a]	15 May 1977	25.87[a]	Intraspecific aggression	Complete
Cusano	M	Mitumba	2 Jul 1956[a]	11 Jun 1996	39.94[a]	Unknown	Nearly complete (post burial damage)
Ebony	M	Mitumba	9 Nov 1996	17 Jan 2005	8.19	Intraspecific aggression	Complete
Echo	F	Kalande	2 Jul 1972[a]	2006	34[a]	Injury	Complete
Flint	M	Kasekela	1 Mar 1964	17 Sep 1972	8.55	Orphaning	Complete
Flo	F	Kasekela	2 Jul 1919[a]	22 Aug 1972	53.14[a]	Illness	Complete
Fred	M	Kasekela	5 Sep 1996	29 July 1997	0.9	Illness	Complete
Gaia's infant	M	Kasekela	16 Jul 2008	23 Jul 2008	0	Unknown	Complete
Galahad	M	Kasekela	5 Apr 1988	7 Feb 2000	11.84	Illness	Complete
Getty	M	Kasekela	21 May 1982	12 Apr 1986	3.89	Poaching	Nearly complete
Gilka	F	Kasekela	2 July 1960	16 May 1979	18.87	Illness	Complete
Goblin	M	Kasekela	6 Sep 1964	24 Aug 2004	39.96	Illness	Complete
Gremlin's baby	M	Kasekela	19 May 1987	23 May 1987	0	Maternal disability	Complete
Groucho	M	Kasekela	1 Jul 1985	12 Oct 1986	1.28	Illness	Complete
Gyre	M	Kasekela	21 Oct 1977	13 Aug 1978	0.81	Illness	complete
Hugo	M	Kasekela	2 Jul 1936[a]	1 Feb 1975	38.58[a]	Illness	Complete
Humphrey	M	Kasekela	2 Jul 1946[a]	6 Jun 1981	34.93[a]	Unknown	Cranium only

(continued)

Table 1.2 (continued)

Chimpanzee	Sex	Community	Birth date	Death date	Age at death (years)	Cause of death	Completeness of remains
Jackson	M	Kasekela	16 Sep 1989	14 Feb 2000	10.41	Illness	Complete
Jomeo	M	Kasekela	2 Jul 1956[a]	16 May 1987	30.87[a]	Illness	Complete
Kidevu?	F	Kasekela	1 June 1966	4 Jan 1992	25.59	Unknown	complete
MacDee	M	Kasekela	2 Jul 1953[a]	15 Dec 1966	13.45[a]	Illness	Complete
Madam Bee	F	Kahama	2 Jul 1947[a]	19 Sep 1975	28.22[a]	Intraspecific aggression	Complete
Mel	M	Kasekela	24 Jan 1984	11 Oct 1994	10.71	Intraspecific aggression	Complete
Melissa	F	Kasekela	2 Jul 1950[a]	24 Oct 1986	36.31[a]	Illness	Complete
Melissa's baby	M	Kasekela	6 Jan 1976	8 Jan 1976	0	Unknown	Complete
Michaelmas	M	Kasekela	7 Oct 1973	3 Oct 1986	12.99	Illness	Complete
Miff	F	Kasekela	2 Jul 1956[a]	28 May 1987	30.9[a]	Illness	Complete
November infanticide	F	Kahama or Kalande	n/a	Nov 1975	0.67[a]	Intraspecific aggression	Complete
October infanticide	M	Kahama or Kalande	n/a	Oct 1975	0.75[a]	Intraspecific aggression	Nearly complete (perimortem trauma)
Old Female	F	n/a	n/a	1974[a]	Old adult	Intraspecific aggression	Complete
Pallas	F	Kasekela	2 Jul 1947[a]	11 Sep 1982	27.2[a]	Illness	Complete
Passion	F	Kasekela	2 Jul 1951[a]	10 Feb 1982	30.61[a]	Illness	Complete
Patti	F	Kasekela	2 July 1961	3 Oct 2005	44.25	Intraspecific aggression	Complete
Patti's infant	n/a	Kasekela	15 Apr 1978	21 Apr 1978	0	Maternal disability	Nearly complete
Plato	M	Kasekela	7 Sep 1970	18 Apr 1973	2.61	Illness	Skull and clavicles only
Rejea	F	Mitumba	22 Dec 1992	30 May 1993	0.27	Intraspecific aggression	Nearly complete (perimortem trauma)
Rix	M	Kasekela	2 Jul 1935[a]	19 Nov 1968	27.38[a]	Injury	Complete

(continued)

Table 1.2 (continued)

Chimpanzee	Sex	Community	Birth date	Death date	Age at death (years)	Cause of death	Completeness of remains
Satan	M	Kasekela	2 Jul 1957[a]	3 May 1987	29.83[a]	Illness	Incomplete
Sherehe	F	Kasekela	25 Jan 1991	5 Nov 2006	15.78	Injury	Complete
Sugar	F	Kasekela	2 Jul 1976	31 May 1987	10.91	Illness	Complete
Tumaini	M	Kalande	1972[a]	13 Aug 2002	30[a]	Poaching	Nearly complete (perimortem trauma)
Unknown male	M	?	n/a	n/a	Adult	Unknown	Cranium only
Vincent	M	Mitumba	2 Jul 1976	22 Dec 2004	28.47	Intraspecific aggression	Complete
Winkle?	F	Kasekela	2 Jul 1959[a]	13 May 1988	28.87[a]	Unknown	Cranium only
Yolanda	F	Kasekela	1983[a]	11 Jul 2007	24[a]	Illness	Complete

? indicates tentative ID
[a]Estimate

The bodies of dead chimpanzees were recovered whenever possible for the purpose of skeletal preservation and for analysis of soft tissue for pathologic lesions (Terio et al. 2011). Skeletons of chimpanzees who died prior to 1987 were cleaned according to protocols laid out elsewhere (Zihlman et al. 1990). Skeletons of chimpanzees who died after 1987 were processed in a different way. Following necropsy (Terio et al. 2011), the bodies were buried in a permeable bag and exhumed after at least 1 year. Skeletons were then carefully cleaned using water and a soft brush and allowed to dry thoroughly.

Because some chimpanzees at Gombe are infected with a strain of the Simian immunodeficiency virus (SIVcpz) (Keele et al. 2009; Santiago et al. 2002), an extra step was added to the cleaning procedure of chimpanzees known to be infected. Infected chimpanzees were briefly (2 min) rinsed in a mild bleach solution after initial cleaning, and then subsequently soaked in water to remove bleach residue. All skeletons were inventoried and stored in mosquito net bags (to reduce the risk of insect damage to the bones in storage) in a secure location managed by the Gombe Stream Research Center.

Dominance rank data for this study were generously made available by Dr. Anne Pusey and have their basis in long-term behavioral research. Rank data are presented in Table 1.3 and are collated from Goodall (1986), Foster et al. (2009), Murray (2007), Murray et al. (2006), and unpublished data from Gombe Stream Research Center (GSRC). Because rank often changes over the course of a chimpanzee's life,

Table 1.3 Dominance rank data for skeletonized chimpanzees from Gombe

Name	Sex	Mother	Age	Highest rank	Rank at death	Mother's rank	Citation
Andromeda	F	Aphro	0.73	3	3	2	Based on age: Pusey et al. (2005), Deus Mjungu (unpublished data)
Atlas	M	Athena	31.3	1	3	2	Wroblewski et al. (2009), GSRC (unpublished data), Goodall (1986)
Beethoven	M	n/a	33.45	1	3	n/a	Wroblewski et al. (2009), GSRC (unpublished data)
Bwavi male	M	n/a	40+	n/a	n/a	n/a	Unknown individual
Charlie	M	n/a	25.87	1	1	n/a	Bygott (1979), Goodall (1986)
Cusano	M	n/a	40.95	1	1	n/a	GSRC (unpublished data)
Ebony	M	Eva	8.39	3	3	n/a	Based on age: Pusey et al. (2005)
Echo	F	n/a	22.5	n/a	n/a	n/a	Data unavailable
Flint	M	Flo	8.55	3	3	1	Based on age: Pusey et al. (2005), Goodall (1986)
Flo	F	n/a	53.14	1	1	n/a	Goodall (1986)
Fred	M	Fifi	0.9	3	3	1	Based on age: Pusey et al. (2005)
Gaia's infant	M	Gaia	0	3	3	2	Based on age: Pusey et al. (2005)
Galahad	M	Gremlin	11.4	3	3	2	Based on age: Pusey et al. (1997, 2005)
Getty	M	Gremlin	3.89	3	3	3	Based on ag:e Pusey et al. (1997, 2005)
Gilka	F	Olly	18.87	3	3	2	Goodall (1986), Pusey et al. (1997)
Gremlin's infant	M	Gremlin	0	3	3	3	Based on age: Pusey et al. (1997, 2005)
Goblin	M	Melissa	39.96	1	1	2	Wroblewski et al. (2009), GSRC (unpublished data)
Groucho	M	Melissa	1.28	3	3	2	Based on age:Pusey et al. (1997, 2005)
Gyre	M	Melissa	0.81	3	3	2	Based on age: Pusey et al. (1997, 2005)
Hugo	M	n/a	38.58	n/a	3	n/a	Goodall (1986)
Humphrey	M	n/a	34.93	1	2	n/a	Goodall (1986)

(continued)

Table 1.3 (continued)

Name	Sex	Mother	Age	Highest rank	Rank at death	Mother's rank	Citation
Nov. Infanticide	F	n/a	0.67	3	3	n/a	Based on age
Oct. Infanticide	M	n/a	0.75	3	3	n/a	Based on age
Jackson	M	Jiffy	10.41	3	3	3	Based on age: Pusey et al. (2005)
Jomeo	M	Vodka	30.87	1	1	n/a	Wroblewski et al. (2009), GSRC (unpublished data)
Kidevu	F	n/a	25.59	1	2	n/a	Data unavailable
MacDee	M	Jessica	13.45	3	3	n/a	Based on age: Pusey et al. (2005)
Madam B	F	n/a	28.22	n/a	n/a	n/a	Data unavailable
Mel	M	Miff	10.71	3	3	1	Based on age: Pusey et al. (1997, 2005)
Melissa	F	n/a	36.31	2	2	n/a	Pusey et al. (1997), Goodall (1986)
Michaelmas	M	Miff	12.99	3	3	2	Based on age: Pusey et al. (1997, 2005)
Miff	F	Marina	30.9	1	1	2	Goodall (1986), Pusey et al. (1997)
Melissa's infant	F	Melissa	0.01	3	3	2	Based on age: Pusey et al. (1997, 2005)
Old Female	F	n/a	Old adult	n/a	n/a	n/a	Data unavailable
Pallas	F	n/a	27.2	3	3	n/a	Pusey et al. (1997)
Passion	F	n/a	30.61	1	1	n/a	Pusey et al. (1997)
Patti	F	n/a	44.25	1	1	n/a	Murray et al. (2006)
Patti's infant	M	Patti	0.02	3	3	3	Based on age: Pusey et al. (1997, 2005)
Plato	M	Pallas	2.61	3	3	3	Based on age: Pusey et al. (1997, 2005)
Rejea	F	Rafiki	0.38	3	3	n/a	Based on age: Pusey et al. (2005)
Rix	M	n/a	27.38	n/a	n/a	n/a	Data unavailable
Satan	M	Sprout	29.83	1	1	n/a	Wroblewski et al. (2009), GSRC (unpublished data)
Sherehe	F	Sandi	15.78	2	2	3	GSRC (unpublished data)
Sugar	F	Caramel	10.91	3	3	n/a	Based on age: Pusey et al. (2005)
Tumaini	M	n/a	30	n/a	n/a	n/a	Data unavailable

(continued)

Table 1.3 (continued)

Name	Sex	Mother	Age	Highest rank	Rank at death	Mother's rank	Citation
Unknown male	M	n/a	Adult	n/a	n/a	n/a	Data unavailable
Vincent	M	n/a	28.47	1	3	n/a	GSRC (unpublished data)
Winkle	F	n/a	28.9	1	1	n/a	Pusey et al. (1997)
Yolanda	F	n/a	24.35	2	2	n/a	GSRC (unpublished data)

1 = high ranking, 2 middle ranking, 3 = low ranking

Table 1.4 Scoring system for pathology and trauma (after Carter et al. (2008))

Pathology/trauma	Elements affected	Extent of affected area	Severity	Degree of healing
1. Trauma	e.g., Humerus	Measured in mm	1. Mild 2. Moderate 3. Severe	1. Perimortem 2. Moderate 3. Advanced
2. Arthropathy				
3. Bone loss				
4. Bone formation				
5. Developmental abnormality				

highest achieved rank and rank at death were both considered whenever possible in an attempt to encompass rank variation over the course of a chimpanzee's life. Maternal rank at the time of a chimpanzee's birth is also included in the table.

Female dominance ranks are usually categorized as high ranking versus middle ranking versus low ranking (e.g., Pusey et al. 2005). Male chimpanzees may be given ordinal ranks (1–*n* adult males in the community) (Bygott 1979; Goodall 1986), but here these were collapsed into categories similar to the females' for the purpose of the statistical analyses undertaken in this study. Males ranked 1–4 are considered high ranking, 5–8 are considered middle ranking, and ranks 9 and above are low ranking. These categories were adjusted twice to account for the number of males in a group. Vincent was ranked third of three adult male chimpanzees at the time of his death, and is therefore considered low ranking. Similarly, Jomeo's numerical rank during the year of his death could be considered either high or middle ranking because of the relatively small number of adult males. Jomeo's relative rank (Jameson et al. 1999) places him in the top third of adult male ranks; however, so he is counted as high ranking at time of death.

Each skeletal element was macroscopically examined under natural light for signs of trauma and pathology. Traumata and pathologies were scored according to the system outlined by Carter et al. (2008) in a study of chimpanzee skeletons from Kibale National Park, Uganda. Table 1.4 outlines the categories used in this study. To the five categories outlined by Carter and colleagues, I added a sixth category: "indeterminate/combined" for those cases when it was unclear what process was at work or for cases where multiple factors accounted equally for the pathology. (These cases make up a very small number (under 1%) of the observed lesions.) This particular scoring system has the advantage of allowing for direct comparison between the Kibale and Gombe skeletal samples. It also provides a structured way to assess pathology and examine trends within the sample without the need for differential diagnoses in each case (which will inherently be more or less accurate/ precise depending on the nature and specificity of the pathology in question). This is a conservative approach to lesion analysis, which relies on frequency and location of lesions much more than descriptions of specific disease processes or sequelae to injuries. Because joint disease can result in both bone loss and bone formation, instances of arthropathy were only scored in this category and not also scored as bone loss or bone formation instances. Arthropathy is considered a separate type of lesion in all instances; more details on the brief examination of joint disease undertaken for this project are presented in Chap. 2. Criteria for determining the category of trauma or pathology followed protocols laid out in bone disease manuals (Aufderheide and Rodriguez-Martín 1998; Ortner 2003; Rogers and Waldron 1995; Buikstra and Ubelaker 1994; Resnick 1988).

Dental eruption, cranial suture fusion, and post-cranial epiphyseal fusion were documented using categories laid out in Tables 1.5, 1.6, and 1.7. Results for individual chimpanzees are documented in Tables 1.8, 1.9, 1.10, 1.11, 1.12, 1.13, 1.14, 1.15, 1.16, 1.17, 1.18, 1.19, 1.20, 1.21, and 1.22.

Many of the data on trauma and pathology (Jurmain 1989, 1997, 2000; Jurmain and Kilgore 1998; Kilgore 1989) as well as skeletal and dental maturation (Zihlman et al. 1990, 2004b) have been previously published for chimpanzees who died prior to 1987, but I summarize them again here using a slightly finer-scale scoring system for consistency. Ages and sexes are known from the behavioral literature (e.g., Goodall 1986) when positive identification of an individual at the time of death was possible. Unknown individuals, individuals of uncertain identification, and individuals with estimated ages are indicated in Table 1.2.

Table 1.5 Dental eruption scoring system

Score	Description
0	Shed deciduous tooth
1	Unerupted, not visible
2	Tooth unerupted, but visible in crypt
3	Tooth is less than half erupted
4	Tooth is more than half erupted
5	Tooth is in full occlusion
6	Antemortem tooth loss of permanent tooth
7	Postmortem tooth loss or severe damage
8	Congenitally absent

Table 1.6 Scoring system for cranial suture fusion

Score	Description
x	Unobservable
0	Open
1	Straight
2	Interdigitated
3	Suture less than 50% obliterated
4	Suture more than 50% obliterated
5	Suture totally obliterated

Scores after Buikstra and Ubelaker (1994), but with additional categories to reflect inclusion of sub-adults with developing crania in the sample

Table 1.7 Scoring system for post-cranial epiphyseal fusion

Score	Description
1	Epiphysis has not begun to fuse (totally unattached to the diaphysis)
2	The epiphysis is fused to the diaphysis around half or less of the circumference of the shaft
3	The epiphysis is fused around more than half of the circumference of the shaft
4	The epiphysis is fused around the entire shaft, but a line or gap is still visible along some portion of the epiphysis
5	The epiphyseal line is completely obliterated

1.5 Individual Cases

The narratives below are meant to provide an overall picture of the traumata and pathologies accumulated on each individual's skeleton at the time of death rather than complete descriptions of each affected element.

Melissa's Infant (1976) (Male: 2 Days)
This is the mostly complete skeleton of a chimpanzee neonate. He was Melissa's fifth recorded pregnancy; Melissa is estimated to have been 26 years old when this infant was born (Goodall 1986). All epiphyses are unfused, no secondary ossification centers are evident, and no teeth are erupted. The developing crowns of 14 deciduous teeth are associated with the skeleton, but are no longer in situ. There are 15 metacarpals/metatarsals associated with the skeleton, but no carpals or tarsals. The age of this infant is recorded as 2 days in unpublished data from Gombe Stream Research Center (GSRC), though this infant has previously been described as 1 week old at death (Goodall 1986).

The infant is suspected to have been killed by Passion and Pom (Goodall 1986), but it seems likely Melissa was able to retain possession of her son, as she carried

Table 1.8 Dental eruption scores for the deciduous dentition of individual sub-adult chimpanzees

Specimen	Sex	Age	A	B	C	D	E	F	G	H	I	J	K	L	M	N	O	P	Q	R	S	T	Notes
Gaia's infant	M	0	1	1	1	1	1	1	1	1	1	1	1	1	1	1	1	1	1	1	1	1	13 tooth buds
Gremlin's infant	M	0	1	1	1	1	1	1	1	1	1	1	1	1	1	1	1	1	1	1	1	1	
Melissa's infant	M	0	1	1	1	1	1	1	1	1	1	1	1	1	1	1	1	1	1	1	1	1	
Patti's infant	?	0	1	1	1	1	1	1	1	1	1	1	1	1	1	1	1	1	1	1	1	1	
Rejea	F	0.27	2	5	1	5	5	5	5	1	5	2	3	5	1	5	5	5	5	1	5	3	
Nov. Infanticide	F	0.67[a]	3	5	2	5	5	5	5	2	5	3	4	5	2	5	5	5	5	2	5	4	
Andromeda	F	0.73	5	5	5	5	5	5	5	5	5	5	5	5	5	5	5	5	5	5	5	5	
Oct. Infanticide	M	0.75[a]	5	7	7	7	7	7	7	7	7	5	5	5	3	5	5	5	5	3	5	5	
Gyre	M	0.81	3	5	2	3	5	5	3	2	5	3	3	7	2	3	4	4	3	2	7	3	
Fred	M	0.9	4	5	3	5	5	5	5	3	5	4	4	5	3	5	5	5	5	3	5	4	
Groucho	M	1.28	5	5	3	4	5	5	4	3	5	5	5	5	3	5	5	5	5	3	5	5	
Plato	M	2.61	5	5	5	5	5	5	5	5	5	5	5	5	5	5	5	5	5	5	5	5	
Getty	M	3.89																					No skull
Ebony	M	8.19	0	5	0	0	0	0	0	5	0	0	0	0	5	0	0	0	0	5	0	0	
Flint	M	8.55	5	5	5	0	0	0	0	5	5	5	5	5	5	0	0	0	0	5	5	5	
Jackson	M	10.41	5	5	5	0	0	0	0	5	5	5	0	0	5	0	0	0	0	5	5	5	

Mel	M	10.71	0	0	5	0	0	0	0	5	0	0	0	0	0	0	0	0	0	0	0	0	Left mandibular deciduous canine shed, alveolus absent, no adult tooth erupting
Sugar	F	10.91	0	0	0	0	0	0	0	0	0	0	0	0	0	0	0	0	0	0	0	0	Only left maxilla preserved
Galahad	M	11.84	0	0	0	0	0	0	0	0	0	0	0	0	0	0	0	0	0	0	0	0	Most teeth lost postmortem, but were likely in full occlusion based on alveolar development
Michaelmas	M	12.99	0	0	0	0	0	0	0	0	0	0	0	0	0	0	0	0	0	0	0	0	
MacDee	M	13.45[a]	0	0	0	0	0	0	0	0	0	0	0	0	0	0	0	0	0	0	0	0	

Universal dental numbering system employed, with letters for deciduous dentition and Arabic numerals for the permanent dentition

[a]Indicates estimated age

Table 1.9 Dental eruption scores for the permanent maxillary dentition of individual sub-adult chimpanzees

Specimen	Sex	Age	1	2	3	4	5	6	7	8	9	10	11	12	13	14	15	16	Notes
Gaia's infant	M	0	1	1	1	1	1	1	1	1	1	1	1	1	1	1	1	1	13 tooth buds
Gremlin's infant	M	0	1	1	1	1	1	1	1	1	1	1	1	1	1	1	1	1	
Melissa's infant	M	0	1	1	1	1	1	1	1	1	1	1	1	1	1	1	1	1	
Patti's infant	?	0	1	1	1	1	1	1	1	1	1	1	1	1	1	1	1	1	
Rejea	F	0.27	1	1	2	1	1	1	1	1	1	1	1	1	1	2	1	1	
Nov. Infanticide	F	0.67[a]	1	1	2	1	1	1	1	1	1	1	1	1	1	2	1	1	
Andromeda	F	0.73	1	1	2	1	1	1	1	1	1	1	1	1	1	2	1	1	
Oct. Infanticide	M	0.75[a]	1	1	2	1	1	1	1	1	1	1	1	1	1	2	1	1	
Gyre	M	0.81	1	1	2	1	1	1	1	1	1	1	1	1	1	2	1	1	
Fred	M	0.9	0	0	2	0	0	0	0	0	0	0	0	0	0	2	0	0	
Groucho	M	1.28	1	1	2	1	1	1	2	2	2	2	1	1	1	1	1	1	
Plato	M	2.61	1	1	2	1	1	1	2	2	2	2	1	1	1	2	1	1	
Getty	M	3.89																	No skull
Ebony	M	8.19	2	5	5	5	4	2	5	5	5	5	2	4	5	5	5	2	
Flint	M	8.55	1	4	5	1	1	1	5	5	5	5	1	1	1	5	4	1	
Jackson	M	10.41	2	5	5	1	1	1	5	5	5	5	1	1	1	5	5	2	
Mel	M	10.71	2	5	5	5	5	2	5	5	5	5	2	5	5	5	5	2	Left mandibular deciduous canine shed, alveolus absent, no adult tooth erupting
Sugar	F	10.91	7	7	7	7	7	7	7	7	5	5	2	5	5	5	5	2	Only left maxilla preserved
Galahad	M	11.84	7	7	7	7	7	4	5	5	5	5	4	7	7	7	7	7	Most teeth lost postmortem, but were likely in full occlusion based on alveolar development
Michaelmas	M	12.99	5	5	5	5	5	4	5	5	5	5	4	5	5	5	5	5	
MacDee	M	13.45[a]	5	5	5	5	5	5	5	5	5	5	5	5	5	5	5	5	

Universal dental numbering system employed, with letters for deciduous dentition and Arabic numerals for the permanent dentition

[a]Indicates estimated age

Table 1.10 Dental eruption scores for the permanent mandibular dentition of individual sub-adult chimpanzees

Specimen	Sex	Age	17	18	19	20	21	22	23	24	25	26	27	28	29	30	31	32	Notes
Gaia's infant	M	0	1	1	1	1	1	1	1	1	1	1	1	1	1	1	1	1	13 tooth buds
Gremlin's infant	M	0	1	1	1	1	1	1	1	1	1	1	1	1	1	1	1	1	
Melissa's infant	M	0	1	1	1	1	1	1	1	1	1	1	1	1	1	1	1	1	
Patti's baby	?	0	1	1	1	1	1	1	1	1	1	1	1	1	1	1	1	1	
Rejea	F	0.27	1	1	2	1	1	1	1	1	1	1	1	1	1	2	1	1	
Nov. Infanticide	F	0.67[a]	1	1	2	1	1	1	1	1	1	1	1	1	1	2	1	1	
Andromeda	F	0.73	1	1	2	1	1	1	1	1	1	1	1	1	1	2	1	1	
Oct. Infanticide	M	0.75[a]	1	1	2	1	1	1	1	1	1	1	1	1	1	2	1	1	
Gyre	M	0.81	1	1	2	1	1	1	1	1	1	1	1	1	1	2	1	1	
Fred	M	0.9	0	0	2	0	0	0	0	0	0	0	0	0	0	2	0	0	
Groucho	M	1.28	1	1	2	1	1	1	1	1	1	1	1	1	1	2	1	1	
Plato	M	2.61	1	2	2	1	1	1	1	1	1	1	1	1	1	2	2	1	
Getty	M	3.89																	No skull
Ebony	M	8.19	2	5	5	5	5	2	5	5	5	5	2	5	5	5	5	5	
Flint	M	8.55	1	4	5	1	2	1	5	5	5	5	1	2	1	5	4	1	
Jackson	M	10.41	2	5	5	2	5	2	5	5	5	5	2	2	1	5	5	2	

(continued)

Table 1.10 (continued)

Specimen	Sex	Age	17	18	19	20	21	22	23	24	25	26	27	28	29	30	31	32	Notes
Mel	M	10.71	2	5	5	5	5	8?	5	5	5	5	3	5	5	5	5	2	Left mandibular deciduous canine shed, alveolus absent, no adult tooth erupting
Sugar	F	10.91	2	5	5	5	5	3	5	5	5	5	3	5	5	5	5	2	Only left maxilla preserved
Galahad	M	11.84	7	7	7	7	7	4	5	7	7	5	4	7	7	7	7	7	Most teeth lost postmortem, but were likely in full occlusion based on alveolar development
Michaelmas	M	12.99	5	5	5	5	5	5	5	5	5	5	5	5	5	5	5	5	
MacDee	M	13.45[a]	5	5	5	5	5	5	5	5	5	5	5	5	5	5	5	5	

Universal dental numbering system employed, with letters for deciduous dentition and Arabic numerals for the permanent dentition
[a]Indicates estimated age

Table 1.11 Dental eruption scores for the permanent maxillary dentition of individual adult chimpanzees

Specimen	Sex	Age	1	2	3	4	5	6	7	8	9	10	11	12	13	14	15	16	Notes
Sherehe	F	15.78	5	5	5	5	5	5	5	7	5	5	5	5	5	5	5	5	
Gilka	F	18.87	5	5	5	5	5	5	5	5	5	5	5	5	5	5	5	5	
Yolanda	F	24[a]	5	5	5	5	5	5	5	5	5	5	5	5	5	5	5	5	
Kidevu	F	25.59	5	5	5	5	5	5	5	7	5	5	5	5	5	5	5	5	Fourth upper left molar fully erupted
Charlie	M	25.87[a]	5	5	5	5	5	5	5	5	5	5	5	5	5	5	5	5	
Pallas	F	27.2[a]	5	5	5	5	5	5	5	5	5	7	5	5	5	5	5	5	
Rix	M	27.38[a]	5	5	5	5	5	5	5	5	5	5	5	5	5	5	5	5	
Madam Bee	F	28.22[a]	5	5	5	5	5	5	5	5	5	5	5	5	5	5	5	5	
Vincent	M	28.47	5	5	5	5	5	5	5	5	5	5	5	5	5	5	5	5	
Winkle	F	28.87[a]	No maxilla or mandible																
Satan	M	29.83[a]	No cranium																
Tumaine	M	30[a]	5	5	5	5	5	5	5	5	5	5	5	5	5	5	5	5	
Passion	F	30.61[a]	5	5	5	5	5	5	5	5	5	5	5	5	5	5	5	5	
Jomeo	M	30.87[a]	5	5	5	5	5	5	5	5	5	5	5	5	5	5	5	5	Only root stump remains for upper left canine
Miff	F	30.9[a]	5	5	5	5	5	5	5	5	5	5	5	6	5	7	6	7	
Atlas	M	31.3[a]	5	5	5	5	5	5	5	5	5	5	5	6	5	5	5	5	

(continued)

Table 1.11 (continued)

Specimen	Sex	Age	1	2	3	4	5	6	7	8	9	10	11	12	13	14	15	16	Notes
Beethoven	M	33.45[a]	5	5	6	5	5	5	5	6	5	5	5	5	5	5	5	5	
Echo	F	34[a]	5	5	5	5	5	5	5	5	5	5	5	5	5	5	5	5	
Humphrey	M	34.93[a]	5	5	5	5	5	5	5	5	5	5	5	5	5	5	5	5	
Melissa	F	36.31[a]	5	5	5	5	5	5	5	5	5	5	5	5	5	5	5	5	
Hugo	M	38.58[a]	5	5	5	5	6	5	5	5	5	5	7	6	5	5	5	5	
Cusano	M	39.94[a]	5	5	5	5	5	5	5	5	5	5	5	5	5	5	5	5	
Goblin	M	39.96	5	5	5	5	5	5	5	5	5	5	5	5	5	5	5	5	
Bwavi male	M	40 + ?[a]	5	5	5	5	5	5	5	5	5	5	5	5	5	5	5	5	
Patti	F	44.25	5	5	5	5	5	5	5	5	5	5	5	5	5	6	5	5	
Flo	f	53.14[a]	5	5	5	5	5	5	6	6	6	6	5	5	5	5	5	5	Many remaining teeth worn to root stumps. Alveolar remodeling used to determine AMTL.
Unknown Mitumba M	M	Adult	7	5	7	7	7	7	7	7	7	7	7	7	7	7	5	7	
Old female	F	Old adult	5	5	5	5	7	5	6	6	6	6	5	7	5	5	7	5	

Universal dental numbering system employed
[a]Indicates estimated age

Table 1.12 Dental eruption scores for the permanent mandibular dentition of individual adult chimpanzees

Specimen	Sex	Age	17	18	19	20	21	22	23	24	25	26	27	28	29	30	31	32	Notes
Sherehe	F	15.78	5	5	5	5	5	5	5	5	5	5	5	5	5	5	5	5	
Gilka	F	18.87	5	5	5	5	5	5	5	5	5	5	5	5	5	5	5	5	
Yolanda	F	24[a]	5	5	5	5	5	5	5	5	5	5	5	5	5	5	5	5	
Kidevu	F	25.59	5	5	5	5	5	5	5	5	5	5	5	5	5	5	5	5	
Charlie	M	25.87[a]	5	5	5	5	5	5	5	5	5	5	5	5	5	5	5	5	
Pallas	F	27.2[a]	5	5	5	5	5	5	5	5	5	5	5	5	5	5	5	5	
Rix	M	27.38[a]	7	5	5	5	5	5	5	5	5	5	5	5	5	5	5	5	
Madam Bee	F	28.22[a]	5	5	5	5	5	5	5	5	5	5	5	5	5	5	5	5	
Vincent	M	28.47	5	5	6	6	6	6	6	6	6	6	5	5	5	5	5	5	
Winkle	f	28.87[a]																	No maxilla or mandible
Satan	M	29.83[a]	7	7	5	5	5	5	7	7	7	5	5	5	5	5	5	7	
Tumaine	M	30[a]	5	5	5	5	5	5	5	5	5	5	5	5	5	5	5	5	
Passion	F	30.61[a]	5	5	5	5	5	5	5	5	5	5	5	5	5	5	5	5	
Jomeo	M	30.87[a]	5	5	5	6	6	5	5	5	5	5	5	5	5	5	5	5	
Miff	F	30.9[a]	5	5	5	5	5	6	5	5	5	5	5	5	5	5	5	5	
Atlas	M	31.3[a]	5	5	5	5	5	5	5	5	5	5	5	5	5	5	5	5	
Beethoven	M	33.45[a]																	No mandible
Echo	F	34[a]	5	5	5	5	5	5	5	5	5	5	5	5	5	5	5	5	
Humphrey	M	34.93[a]																	No mandible
Melissa	F	36.31[a]	5	5	5	5	5	5	5	5	5	5	5	5	5	5	5	5	
Hugo	M	38.58[a]	6	6	6	6	6	5	5	5	5	5	5	5	6	6	6	6	
Cusano	M	39.94[a]	5	5	5	5	5	5	5	5	5	5	5	5	5	5	5	5	
Goblin	M	39.96	5	5	5	5	5	5	5	5	5	5	5	5	5	5	5	5	

(continued)

Table 1.12 (continued)

Specimen	Sex	Age	17	18	19	20	21	22	23	24	25	26	27	28	29	30	31	32	Notes
Bwavi male	M	40 + ?[a]	5	5	5	5	5	5	5	5	5	5	5	5	5	5	5	5	
Patti	F	44.25	5	5	5	5	5	5	5	5	5	5	5	5	5	5	5	5	
Flo	f	53.14[a]	5	6	6	5	5	5	6	5	5	6	5	6	6	6	6	5	Many remaining teeth worn to root stumps. Alveolar remodeling used to determine AMTL
Unknown Mitumba M	M	Adult																	No mandible
Old female	F	Old adult	7	6	6	7	7	5	5	5	5	5	5	5	7	6	6	7	

Universal dental numbering system employed
[a]Indicates estimated age

Table 1.13 Cranial suture fusion scores for individual sub-adult chimpanzees, superior and anterior locations

Specimen	Sex	Age at death	Metopic suture	Bregma	C1 R	C2 R	C1 L	C2 L	S1	S2	S3	S4	Zygo-temp R	Zygo-temp L	Notes
Gaia's infant	M	0	0	0	0	0	0	0	0	0	0	0	0	0	
Gremlin's infant	M	0	0	0	0	0	0	0	0	0	0	0	0	0	
Melissa's infant	M	0	0	0	0	0	0	0	0	0	0	0	0	0	
Patti's infant	?	0	0	0	0	0	0	0	0	0	0	0	0	0	
Rejea	F	0.27	4	x	1	1	x	x	1	1	1	1	x	x	
Nov. Infanticide	F	0.67[a]	3	1	1	1	1	1	1	1	1	1	1	1	
Oct. Infanticide	M	0.75[a]	3	1	1	1	1	1	1	1	1	1	1	1	
Andromeda	F	0.73	4	0	1	1	1	1	1	1	1	1	1	1	
Gyre	M	0.81	0	0	0	0	0	0	0	0	0	0	0	0	
Fred	M	0.9	4	1	1	1	1	1	1	2	2	1	1	1	
Groucho	M	1.28	4	2	2	1	2	1	2	2	2	2	0	0	Unfused mandibular symphysis + AR42
Plato	M	2.61	4	2	2	2	2	2	2	2	2	2	2	1	
Getty	M	3.89	No skull												
Ebony	M	8.19	5	2	2	2	2	2	2	2	2	2	2	2	
Flint	M	8.55	4	2	2	2	2	2	2	2	2	2	2	2	
Jackson	M	10.41	5	2	2	2	2	2	2	2	2	2	1	2	
Mel	M	10.71	4	2	2	1	2	1	2	2	2	2	1	1	
Sugar	F	10.91	4	2	2	2	2	2	2	2	2	2	x	x	No R max; L max not attached to neurocranium; rest of splanchnocranium gone
Galahad	M	11.84	5	2	2	2	2	2	2	2	2	2	2	2	

(continued)

Table 1.13 (continued)

Specimen	Sex	Age at death	Metopic suture	Bregma	C1 R	C2 R	C1 L	C2 L	S1	S2	S3	S4	Zygo-temp R	Zygo-temp L	Notes
Michaelmas	M	12.99	5	2	2	2	2	2	2	2	2	2	2	2	
MacDee	M	13.45[a]	5	2	2	3	2	3	3	4	4	4	2	2	

Ectocranial suture locations after Key et al. (1994)

[a]Indicates estimated age

Table 1.14 Cranial suture fusion scores for individual sub-adult chimpanzees, lateral and posterior locations

Specimen	Sex	Age at death	Lambda	L1 R	L2 R	L1 L	L2 L	OC1 R	OC2 R	OC3 R	OC1 L	OC2 L	OC3 L	SQ1 R	SQ2 R	SQ1 L	SQ2 L	Pterion R	Pterion L	SpT1 R	SpT2 R	SpT L	SpT2 L	SpF R	SpF L	Notes
Gaia's infant	M	0	0	0	0	0	0	0	0	0	0	0	0	0	0	0	0	0	0	0	0	0	0	0	0	
Gremlin's infnt	M	0	0	0	0	0	0	0	0	0	0	0	0	0	0	0	0	0	0	0	0	0	0	0	0	
Melissa's infant	M	0	0	0	0	0	0	0	0	0	0	0	0	0	0	0	0	0	0	0	0	0	0	0	0	
Patti's infant	?	0	0	0	0	0	0	0	0	0	0	0	0	0	0	0	0	0	0	0	0	0	0	0	0	
Rejea	f	0.27	x	1	1	1	1	1	1	1	1	1	1	1	1	x	x	x	x	x	x	x	x	x	x	
Nov. Infanticide	F	0.67[a]	1	1	1	1	1	1	1	1	1	1	1	1	1	1	1	1	1	1	1	1	1	1	1	
Oct. Infanticide	M	0.75[a]	1	1	1	1	1	1	1	1	1	1	1	1	1	1	1	1	1	1	1	1	1	1	1	
Andromeda	F	0.73	1	1	1	1	1	1	1	1	1	1	1	1	1	1	1	1	1	1	1	1	1	1	1	
Gyre	M	0.81	0	0	0	0	0	0	0	0	0	0	0	0	0	0	0	0	0	0	0	0	0	0	0	
Fred	M	0.9	1	1	1	1	1	1	1	1	1	1	1	1	1	1	1	1	1	1	1	1		1	1	
Groucho	m	1.28	0	1	1	1	1	0	0	0	0	0	0	2	1	2	1	0	0	0	0	0	0	0	0	Unfused mandibular symphysis + AR42
Plato	M	2.61	2	2	2	2	2	2	1	1	2	1	1	2	1	2	1	2	2	2	2	2	2	1	2	
Getty	M	3.89	No skull																							
Ebony	M	8.19	2	2	2	2	2	2	3	1	1	1	1	2	1	2	1	1	2	2	1	2	3	3	3	
Flint	M	8.55	2	2	2	2	2	1	1	1	1	1	1	1	1	1	1	2	2	2	2	2	2	2	2	

(continued)

Table 1.14 (continued)

Specimen	Sex	Age at death	Lambda	L1 R	L2 R	L1 L	L2 L	OC1 R	OC2 R	OC3 R	OC1 L	OC2 L	OC3 L	SQ1 R	SQ2 R	SQ1 L	SQ2 L	Pterion R	Pterion L	SpT1 R	SpT2 R	SpT L	SpT2 L	SpF R	SpF L	Notes
Jackson	M	10.41	2	2	2	2	2	1	1	1	1	1	1	1	1	1	1	1	1	1	1	1	1	1	1	
Mel	M	10.71	2	2	2	2	2	2	1	1	1	1	1	2	1	2	1	1	1	1	1	1	1	1	1	
Sugar	f	10.91	2	2	2	2	2	1	1	1	1	1	1	1	1	1	1	1	1	2	1	2	1	1	x	No R max; L max not attached to neurocranium; rest of splanchnocranium gone
Galahad	M	11.84	2	2	1	2	1	1	1	1	1	1	1	2	1	2	1`	1	1	2	1	2	1	2	2	
Michaelmas	M	12.99	2	2	2	2	2	1	1	1	1	1	1	2	2	2	2	2	2	2	2	1	2	2	2	
MacDee	M	13.45[a]	5	5	5	5	5	2	1	1	2	1	1	2	2	2	2	2	1	2	2	2	2	2	1	

Ectocranial suture locations after Key et al. (1994)
[a]Indicates estimated age

Table 1.15 Cranial suture fusion scores for individual sub-adult chimpanzees, inferior locations

Specimen	Sex	Age at death	Occipital: squamous-condylar	Occipital: condylar-basilar	Spheno-occipital synchondrosis	OC1 R	OC2 R	OC3 R	OC1 L	OC2 L	OC3 L	Palatine	Mid max	Incisive	Notes
Gaia's infant	M	0	0	0	0	0	0	0	0	0	0	0	0	0	
Gremlin's infant	M	0	0	0	0	0	0	0	0	0	0	0	0	0	
Melissa's infant	M	0	0	0	0	0	0	0	0	0	0	0	0	0	
Patti's infant	?	0	0	0	0	0	0	0	0	0	0	0	0	0	
Rejea	F	0.27	1	1	1	1	1	1	1	1	1	1	1	1	
Nov. Infanticide	f	0.67[a]	1	1	1	1	1	1	1	1	1	1	1	1	
Oct. Infanticide	M	0.75[a]	1	1	1	1	1	1	1	1	1	x	x	x	
Andromeda	F	0.73	0	0	0	1	1	1	1	1	1	1	1	3	
Gyre	M	0.81	0	0	0	0	0	0	0	0	0	0	0	0	
Fred	M	0.9	1	1	1	1	1	1	1	1	1	1	1	1	
Groucho	M	1.28	0	0	0	0	0	0	0	0	0	2	1	2	Unfused mandibular symphysis + AR42
Plato	M	2.61	2	3	1	2	1	1	2	1	1	2	2	2	
Getty	M	3.89	No skull												
Ebony	M	8.19	5	5	1	2	3	1	1	1	1	2	2	3	
Flint	M	8.55	5	5	0	1	1	1	1	1	1	2	2	2	
Jackson	M	10.41	5	5	1	1	1	1	1	1	1	2	1	3	

(continued)

Table 1.15 (continued)

Specimen	Sex	Age at death	Occipital: squamous-condylar	Occipital: condylar-basilar	Spheno-occipital synchondrosis	OC1 R	OC2 R	OC3 R	OC1 L	OC2 L	OC3 L	Palatine	Mid max	Incisive	Notes
Mel	m	10.71	5	5	1	2	1	1	1	1	1	2	1	3	
Sugar	F	10.91	5	5	1	1	1	1	1	1	1	3	x	3	No R max; L max not attached to neurocranium; rest of splanchnocranium gone
Galahad	M	11.84	5	5	1	1	1	1	1	1	1	2	1	3	
Michaelmas	M	12.99	5	5	0	1	1	1	1	1	1	2	2	1	
MacDee	M	13.45[a]	5	5	3	2	1	1	2	1	1	2	2	2	

Ectocranial suture locations after Key et al. (1994)
[a]Indicates estimated age

Table 1.16 Cranial suture fusion scores for individual adult chimpanzees, superior and anterior locations

Specimen	Sex	Age at death	Metopic suture	Bregma	C1 R	C2 R	C1 L	C2 L	S1	S2	S3	S4	Zygo-temp R	Zygo-temp L	Notes
Sherehe	F	15.78	5	2	2	2	2	2	2	2	3	3	2	2	
Yolanda	F	24[a]	5	5	5	5	5	5	5	5	5	5	5	4	
Gilka	F	18.87	4	2	2	2	2	2	3	5	5	5	4	3	
Kidevu	F	25.59	5	3	3	2	2	2	5	5	5	5	1	2	
Charlie	M	25.87[a]	5	3	3	2	3	2	4	4	4	4	2	2	
Pallas	F	27.2[a]	5	5	5	5	5	5	5	5	5	5	3	3	
Rix	M	27.38[a]	5	5	5	5	5	5	5	5	5	5	4	4	
Madam Bee	F	28.22[a]	5	5	5	5	5	5	5	5	5	5	3	3	
Vincent	M	28.47	5	5	5	5	5	5	5	5	5	5	2	5	
Winkle	M	28.87[a]	5	4	4	5	4	5	5	5	5	5	x	x	Cranium only—no face
Satan	M	29.83[a]	No cranium												
Tumaine	M	30[a]	5	3	3	1	4	4	4	4	4	2	2	2	
Passion	F	30.61[a]	5	5	5	5	5	5	5	5	5	5	5	4	
Jomeo	M	30.87[a]	5	5	5	5	5	5	5	5	5	5	2	2	
Miff	F	30.9[a]	5	5	5		5	5	5	5	5	5	2	x	
Atlas	M	31.3[a]	5	5	5	5	5	5	5	5	5	5	4	3	
Beethoven	M	33.45[a]	5	2	2	2	2	2	1	2	2	2	2	2	Cranium only
Echo	F	34[a]	5	5	5	5	5	5	5	5	5	5	5	5	
Humphrey	M	34.93[a]	5	5	5	5	5	5	5	5	5	5	x	x	

(continued)

Table 1.16 (continued)

Specimen	Sex	Age at death	Metopic suture	Bregma	C1 R	C2 R	C1 L	C2 L	S1	S2	S3	S4	Zygo-temp R	Zygo-temp L	Notes
Melissa	F	36.31[a]	5	5	5	5	5	5	5	5	5	5	5	5	
Hugo	M	38.58[a]	5	5	5	5	5	5	5	5	5	5	3	3	
Goblin	M	39.96	5	5	5	5	5	5	5	5	5	5	3	3	
Cusano	M	39.94[a]	5	5	5	5	5	5	5	5	5	5	x	5	
Bwavi male	M	40 + ?[a]	5	5	5	5	5	5	5	5	5	5	4	4	
Patti	F	44.25	5	5	5	5	5	5	5	5	5	5	5	5	
Flo	F	53.14[a]	5	5	5	5	5	5	5	5	5	5	5	5	
Unknown Mitumba M	M	Adult	4	2	2	2	2	2	2	5	5	4	2	3	Cranium only
Old female	F	Old adult	5	5	5	5	5	5	5	5	5	5	5	5	

Ectocranial suture locations after Key et al. (1994)

[a]Indicates estimated age

Table 1.17 Cranial suture fusion scores for individual adult chimpanzees, lateral and posterior locations

Specimen	Sex	Age at death	Lambda	L1 R	L2 R	L1 L	L2 L	SQ1 R	SQ2 R	SQ1 L	SQ2 L	Pterion R	Pterion L	SpT1 R	SpT2 R	SpT L	SpT2 L	SpF R	SpF L	Notes
Sherehe	F	15.78	4	4	4	4	4	1	1	1	1	3	2	2	2	2	2	3	3	
Yolanda	F	24[a]	5	5	5	5	5	5	5	3	3	5	5	4	5	4	5	5	5	
Gilka	F	18.87	2	2	2	2	2	2	2	2	2	2	2	2	2	2	2	5	5	
Kidevu	F	25.59	2	2	2	2	2	1	2	2	2	5	5	2	2	3	3	5	4	
Charlie	M	25.87[a]	4	3	3	3	3	1	1	1	1	1	1	4	4	5	4	5	5	
Pallas	F	27.2[a]	5	5	5	5	5	4	5	3	5	5	5	5	5	5	5	5	5	
Rix	M	27.38[a]	5	5	5	5	5	4	4	4	4	4	4	4	4	4	4	4	4	
Madam Bee	F	28.22[a]	5	5	5	5	4	4	5	3	5	5	5	5	5	5	5	5	5	
Vincent	M	28.47	5	5	5	5	5	5	5	5	5	5	5	5	5	5	5	5	5	
Winkle	M	28.87[a]	4	4	4	4	4	3	3	5	3	3	5	2	2	3	5	x	5	Cranium only—no face
Satan	M	29.83[a]	No cranium																	
Tumaine	M	30[a]	2	2	2	2	2	1	2	1	2	1	1	2	1	1	1	2	2	
Passion	F	30.61[a]	5	5	5	5	5	3	5	3	5	5	5	5	5	5	5	5	5	
Jomeo	M	30.87[a]	5	5	5	5	5	2	2	2	2	2	2	2	2	2	2	2	2	
Miff	F	30.9[a]	5	5	5	5	5	2	2	2	2	2	3	3	4	2	4	5	5	
Atlas	M	31.3[a]	5	5	5	5	5	3	4	3	3	3	3	3	3	3	3	3	3	
Beethoven	M	33.45[a]	1	1	2	1	2	3	1	1	1	2	2	2	1	2	1	3	3	Cranium only
Echo	F	34[a]	5	5	5	5	3	5	5	5	5	5	5	5	5	5	5	5	5	
Humphrey	M	34.93[a]	2	2	3	2	2	3	3	1	1	5	5	3	2	5	5	5	5	
Melissa	F	36.31[a]	5	5	5	5	5	5	5	5	5	5	5	5	5	5	5	5	5	

(continued)

Table 1.17 (continued)

Specimen	Sex	Age at death	Lambda	L1 R	L2 R	L1 L	L2 L	SQ1 R	SQ2 R	SQ1 L	SQ2 L	Pterion R	Pterion L	SpT1 R	SpT2 R	SpT L	SpT2 L	SpF R	SpF L	Notes
Hugo	M	38.58[a]	5	5	5	5	5	2	2	2	2	2	2	2	1	1	1	2	2	
Goblin	M	39.96	5	5	5	5	5	3	3	4	4	5	5	3	3	4	3	5	5	
Cusano	M	39.94[a]	5	5	5	5	5	5	5	5	5	5	5	5	5	5	5	5	5	
Bwavi male	M	40 + ?[a]	5	5	5	5	5	4	5	4	5	5	5		5	5	5	5	5	
Patti	F	44.25	5	5	5	5	5	5	5	5	5	5	5	4	5	5	5	5	5	
Flo	F	53.14[a]	5	5	5	5	5	5	5	5	5	5	5	5	5	5	5	5	5	
Unknown Mitumba M	M	Adult	2	2	2	2	2	2	2	2	2	2	2	2	2	2	1	1	4	Cranium only
Old female	F	Old adult	3	3	3	3	3	5	5	5	5	5	5	5	5	5	5	5	5	

Ectocranial suture locations after Key et al. (1994)

[a]Indicates estimated age

Table 1.18 Cranial suture fusion scores for individual adult chimpanzees, inferior locations

Specimen	Sex	Age at death	Occipital: squamous-condylar	Occipital: condylar-basilar	Spheno-occipital synchondrosis	OC1 R	OC2 R	OC3 R	OC1 L	OC2 L	OC3 L	Palatine	Mid max	Incisive	Notes
Sherehe	F	15.78	5	5	3	5	4	3	5	3	3	3	4	4	
Yolanda	F	24[a]	5	5	5	2	2	2	2	3	3	5	3	3	
Gilka	F	18.87	5	5	5	2	1	1	1	1	1	2	2	4	
Kidevu	F	25.59	5	5	5	2	1	1	1	1	1	3	4	4	
Charlie	M	25.87[a]	5	5	4	2	1	1	2	2	1	4	5	5	
Pallas	F	27.2[a]	5	5	5	5	5	4	5	5	3	5	5	4	
Rix	M	27.38[a]	5	5	5	5	5	5	5	5	5	5	5	5	
Madam Bee	F	28.22[a]	5	5	3	1	3	1	1	2	1	2	5	4	
Vincent	M	28.47	5	5	5	5	2	1	4	4	1	5	5	5	
Winkle	M	28.87[a]	5	5	5	2	1	1	2	1	1	x	x	x	Cranium only—no face
Satan	M	29.83[a]	No cranium												
Tumaine	M	30[a]	5	5	4	1	1	1	1	1	1	4	4	4	
Passion	F	30.61[a]	5	5	5	5	5	4	5	4	3	5	5	4	
Jomeo	M	30.87[a]	5	5	5	3	1	1	5	2	1	5	5	5	
Miff	F	30.9[a]	5	5	5	2	2	1	2	2	1	5	3	3	
Atlas	M	31.3[a]	5	5	5	5	5	4	5	5	4	5	5	5	
Beethoven	M	33.45[a]	5	5	5	1	1	1	1	1	1	5	5	5	Cranium only
Echo	F	34[a]	5	5	5	2	1`	1	2	1	1	5	5	4	
Humphrey	M	34.93[a]	5	5	5	2	1	1	2	1	1	5	5	5	
Melissa	F	36.31[a]	5	5	5	5	5	4	4	5	4	5	5	4	

(continued)

Table 1.18 (continued)

Specimen	Sex	Age at death	Occipital: squamous-condylar	Occipital: condylar-basilar	Spheno-occipital synchondrosis	OC1 R	OC2 R	OC3 R	OC1 L	OC2 L	OC3 L	Palatine	Mid max	Incisive	Notes
Hugo	M	38.58[a]	5	5	5	5	2	1	2	2	1	5	4	4	
Goblin	M	39.96	5	5	5	5	5	4	5	5	4	5	5	5	
Cusano	M	39.94[a]	5	5	5	5	5	3	5	5	3	5	5	5	
Bwavi male	M	40 + ?[a]	5	5	5	5	4	3	3	3	3	5	5	5	
Patti	F	44.25	5	5	5	5	5	5	5	5	5	5	5	5	
Flo	F	53.14[a]	5	5	5	4	4	1	4	1	1	4	5	4	
Unknown Mitumba M	M	Adult	5	5	4	1	1	1	1	1	1	2	3	4	Cranium only
Old female	F	Old adult	5	5	5	3	2	1	3	2	1	5	5	5	

Ectocranial suture locations after Key et al. (1994)
[a]Indicates estimated age

Table 1.19 Post-cranial epiphyseal fusion scores, upper limb, for individual sub-adult chimpanzees

Specimen	Sex	Age	Acromion	Coracoid tip	Coracoid-body	Margin of scap blade	Med clav	Lat clav	Prox hum	Dist hum	Med epicon	Prox rad	Dist rad	Prox uln	Dist uln
Gaia's infant	M	0	1	1	1	1	1	1	1	1	1	1	1	1	1
Gremlin's infant	M	0	1	1	1	1	1	1	1	1	1	1	1	1	1
Melissa's infant	M	0	1	1	1	1	1	1	1	1	1	1	1	1	1
Patti's infant	?	0	1	1	1	1	1	1	1	1	1	1	1	1	1
Rejea	F	0.27	1	1	1	1	1	1	1	1	1	1	1	1	1
Nov. Infanticide	F	0.67[a]	1	1	1	1	1	1	1	1	1	1	1	1	1
Andromeda	F	0.73	1	1	1	1	1	1	1	1	1	1	1	1	1
Oct. Infanticide	M	0.75[a]	1	1	1	1	n/a	1	1	n/a	n/a	n/a	n/a	n/a	n/a
Gyre	M	0.81	1	1	1	1	1	1	1	1	1	1	1	1	1
Fred	M	0.9	1	1	1	1	1	1	1	1	1	1	1	1	1
Groucho	M	1.28	1	1	1	1	1	1	1	1	1	1	1	1	1
Plato	M	2.61	1	1	1	1	1	1	1	1	1	1	1	1	1
Getty	M	3.89	1	1	1	1	1	1	1	1	1	1	1	1	1
Ebony	M	8.19	1	1	1	1	1	1	1	1	1	1	1	1	1
Flint	M	8.55	1	1	1	1	1	1	1	1	1	1	1	1	1
Jackson	M	10.41	1	1	1	1	1	1	1	1	1	1	1	1	1
Mel	M	10.71	1	1	1	1	1	1	1	1	1	1	1	1	1
Sugar	F	10.91	1	1	1	1	1	1	1	1	1	1	1	1	1
Galahad	M	11.84	1	1	1	1	1	1	1	2	1	1	1	1	1
Michaelmas	M	12.99	1	1	3	1	1	1	1	5	3	1	1	2	1
MacDee	M	13.45[a]	1	5	5	1	1	1	1	5	5	5	1	5	1

[a]Indicates estimated age

Table 1.20 Post-cranial epiphyseal fusion scores, lower limb, for individual sub-adult chimpanzees

Specimen	Sex	Age	Iliac crest	Isch-pub-il	Isch-pub	Fem head	Gr troch	Ls troch	Dist fem	Prox tib	Dist tib	Prox fib	Dist fib	Sacrum	Notes
Gaia's infant	M	0	1	1	1	1	1	1	1	1	1	1	1	1	
Gremlin's infant	M	0	1	1	1	1	1	1	1	1	1	1	1	1	
Melissa's infant	M	0	1	1	1	1	1	1	1	1	1	1	1	1	
Patti's infant	?	0	1	1	1	1	1	1	1	1	1	1	1	1	
Rejea	F	0.27	1	1	1	1	1	1	1	1	1	1	1	1	
Nov. Infanticide	f	0.67[a]	1	1	1	1	1	1	1	1	1	1	1	1	
Andromeda	F	0.73	1	1	1	1	1	1	1	1	1	1	1	1	
Oct. Infanticide	M	0.75[a]	1	1	1	1	1	1	n/a	n/a	n/a	n/a	n/a	n/a	All unobservable locations likley = 1
Gyre	M	0.81	1	1	1	1	1	1	1	1	1	1	1	1	
Fred	M	0.9	1	1	1	1	1	1	1	1	1	1	1	1	
Groucho	M	1.28	1	1	1	1	1	1	1	1	1	1	1	1	
Plato	M	2.61	1	1	1	1	1	1	1	1	1	1	1	1	
Getty	M	3.89	1	1	1	1	1	1	1	1	1	1	1	1	
Ebony	M	8.19	1	1	5	1	1	1	1	1	1	1	1	2	
Flint	M	8.55	1	1	3	1	1	1	1	1	1	1	1	2?	
Jackson	M	10.41	1	1	5	1	1	1	1	1	1	1	1	1	
Mel	M	10.71	1	2	5	1	1	1	1	1	1	1	1	1	
Sugar	F	10.91	1	1	5	1	1	1	1	1	1	1	1	1	
Galahad	M	11.84	1	1	5	1	1	1	1	1	1	1	1	1	
Michaelmas	M	12.99	1	3	3	1	1	1	1	1	1	1	1	3	
MacDee	M	13.45[a]	1	5	3	5	5	5	1	3	4	3	4	3	

[a]Indicates estimated age

Table 1.21 Post-cranial epiphyseal fusion scores, upper limb, for individual adult chimpanzees

Specimen	Sex	Age	Acromion	Coracoid tip	Coracoid-body	Margin of scap blade	Med clav	Lat clav	Prox hum	Dist hum	Med epicon	Prox rad	Dist rad	Prox uln	Dist uln
Sherehe	F	15.78	1	5	5	1	1	5	2	5	5	5	3	5	5
Gilka	F	18.87	1	5	5	1	1	5	4	5	5	4	5	5	5
Yolanda	F	24[a]	1	5	5	4	1	5	5	5	5	5	4	5	5
Kidevu	F	25.59	5	5	5	3	1	n/a	5	5	5	5	5	5	5
Charlie	M	25.87[a]	4	5	5	5	3	5	5	5	5	5	5	5	5
Pallas	F	27.2[a]	2	5	5	5	4	5	4	5	5	4	5	5	5
Rix	M	27.38[a]	5	5	5	5	n/a	n/a	5	5	5	5	5	5	5
Madam Bee	F	28.22[a]	3	5	5	3	5	5	5	5	5	5	5	5	5
Vincent	M	28.47	5	5	5	5	4	5	5	5	5	5	5	5	5
Winkle	F	28.87[a]	No postcrania preserved												
Satan	M	29.83[a]	n/a	n/a	n/a	n/a	n/a	n/a	5	5	5	n/a	n/a	5	n/a
Tumaine	M	30[a]	5	5	5	4	n/a	3	4	5	5	4	4	5	4
Passion	F	30.61[a]	5	5	5	3	5	5	5	5	5	5	5	5	5
Jomeo	M	30.87[a]	5	5	5	5	4	5	5	5	5	5	5	5	5
Miff	F	30.9[a]	2	5	5	5	4	5	4	5	5	5	5	5	5
Atlas	M	31.3[a]	5	5	5	5	1	5	4	5	5	4	5	5	5
Beethoven	M	33.45[a]	No postcrania preserved												
Echo	F	34[a]	5	5	5	4	2	5	5	5	5	5	5	5	5
Humphrey	M	34.93[a]	No postcrania preserved												
Melissa	F	36.31[a]	3	5	5	5	5	5	4	5	5	4	5	5	5
Hugo	M	38.58[a]	5	5	5	5	5	5	5	5	5	5	5	5	5

Table 1.21 (continued)

Specimen	Sex	Age	Acromion	Coracoid tip	Coracoid-body	Margin of scap blade	Med clav	Lat clav	Prox hum	Dist hum	Med epicon	Prox rad	Dist rad	Prox uln	Dist uln
Cusano	M	39.94[a]	5	5	5	n/a	1	5	5	5	5	5	5	5	5
Goblin	M	39.96	5	5	5	5	5	5	5	5	5	5	5	5	5
Bwavi male	M	40 + ?[a]	5	5	5	5	5	5	5	5	5	5	5	5	5
Patti	F	44.25	5	5	5	4	5	5	5	5	5	5	5	5	5
Flo	F	53.14[a]	5	5	5	3	5	5	5	5	5	5	5	5	5
Unknown Mitumba M	M	Adult	No postcrania preserved												
Old female	F	Old adult	3	5	5	5	5	5	5	5	5	5	5	5	5

[a]Indicates estimated age

Table 1.22 Post-cranial epiphyseal fusion scores, lower limb, for individual adult chimpanzees

Specimen	Sex	Age	Iliac crest	Isch-pub-il	Isch-pub	Fem head	Gr troch	Ls troch	Dist fem	Prox tib	Dist tib	Prox fib	Dist fib	Sacrum
Sherehe	F	15.78	1	5	5	4	5	5	3	3	4	3	5	3
Gilka	F	18.87	3	5	5	4	5	5	5	5	5	5	5	4
Yolanda	F	24[a]	3	5	5	4	5	5	5	5	5	4	5	5
Kidevu	F	25.59	3	5	5	5	5	5	5	5	5	5	5	4
Charlie	M	25.87[a]	3	5	5	4	4	5	5	5	5	5	5	3
Pallas	F	27.2[a]	5	5	5	5	5	5	5	5	5	5	5	3
Rix	M	27.38[a]	5	5	5	5	5	5	5	5	5	5	5	4
Madam Bee	F	28.22[a]	3	5	5	n/a	n/a	n/a	n/a	n/a	n/a	n/a	n/a	4
Vincent	M	28.47	4	5	5	5	5	5	5	5	5	5	5	4
Winkle	F	28.87[a]		No postcrania preserved										
Satan	M	29.83[a]	4	5	5	n/a	n/a	n/a	n/a	n/a	n/a	n/a	n/a	n/a
Tumaine	M	30[a]	1	5	5	4	5	5	5	4	5	n/a	n/a	4
Passion	F	30.61[a]	4	5	5	5	5	5	5	5	5	5	4	5
Jomeo	M	30.87[a]	6	6	6	6	6	6	6	5	5	5	5	5
Miff	F	30.9[a]	3	5	5	5	5	5	5	5	5	5	5	3
Atlas	M	31.3[a]	3	5	5	5	5	5	5	4	5	5	5	3
Beethoven	M	33.45[a]		No postcrania preserved										
Echo	F	34[a]	4	5	5	5	5	5	5	5	5	5	5	5

(continued)

Table 1.22 (continued)

Specimen	Sex	Age	Iliac crest	Isch-pub-il	Isch-pub	Fem head	Gr troch	Ls troch	Dist fem	Prox tib	Dist tib	Prox fib	Dist fib	Sacrum
Humphrey	M	34.93[a]		No postcrania preserved										
Melissa	F	36.31[a]	5	5	5	4	5	5	5	5	5	4	5	3
Hugo	M	38.58[a]	4	5	5	n/a	n/a	n/a	n/a	5	5	5	4	5
Cusano	M	39.94[a]	5	5	5	5	5	5	5	5	5	5	5	5
Goblin	M	39.96	5	5	5	5	5	5	5	5	5	5	5	5
Bwavi male	M	40 + ?[a]	5	5	5	5	5	5	5	5	5	5	5	4
Patti	F	44.25	5	5	5	5	5	5	5	5	5	5	5	5
Flo	F	53.14[a]	3	5	5	5	5	5	5	5	5	5	5	5
Unknown Mitumba M	M	Adult		No postcrania preserved										
Old female	F	Old adult	5	5	5	5	5	5	5	5	5	5	5	3

[a]Indicates estimated age

him around for some time after he was killed (or fatally injured). Observers noted that the infant's neck seemed broken, that the forehead was open and bleeding, and some skin had been torn from his upper back (Goodall 1986), but there are no skeletal signs of trauma or pathology. Several other chimpanzees in this sample are known to have sustained significant (sometimes fatal) injuries that left no or few signs on the skeleton (e.g., Goblin, Patti, described below). Because Melissa retained possession of her infant, there would have been no opportunity for the infant to be consumed by his possible killers. The skeleton shows no evidence of damage consistent with chimpanzee consumption of infant conspecifics, as is the case for Rejea (a partially consumed infanticide victim described below).

Of interest in this case is the incomplete ossification of cortical bone on some elements, particularly on the maxilla (Fig. 1.1). Because the periosteal layer of the cortex was still forming at the time of death, the trabeculae are easily visible (Scheuer and Black 2004).

Melissa is also part of the Gombe chimpanzee skeletal collection, as are several other relatives of this formerly high-ranking female, including two more of her offspring who did not reach adulthood (Gyre, Groucho), her grandsons Getty and Galahad, an unnamed grandson and great-grandson (next cases), her suspected sibling Humphrey, and her son Goblin, who was once alpha male of the Kasekela community.

Gremlin's Infant (1987) (Male: 4 Days)

This male infant was Gremlin's second recorded pregnancy; he died in 1987 during an illness of his mother's, when she was 17 years old. Gremlin was Melissa's third recorded pregnancy (she was born in 1970, when Melissa is estimated to have been 20 years old). Melissa is also part of the Gombe skeletal collection, but Gremlin is still a member of the Kasekela community (age 48 in November 2018) (Gombe Stream Research Center (GSRC) unpublished data).

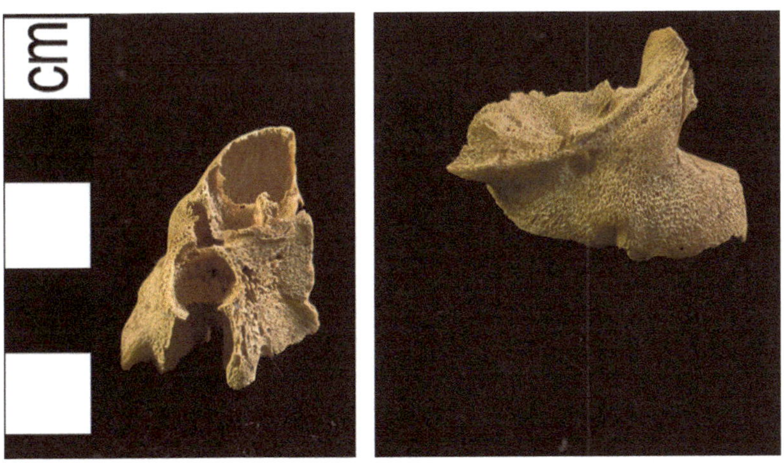

Fig. 1.1 Right maxilla from Melissa's infant: palatal view (left), lateral view (right)

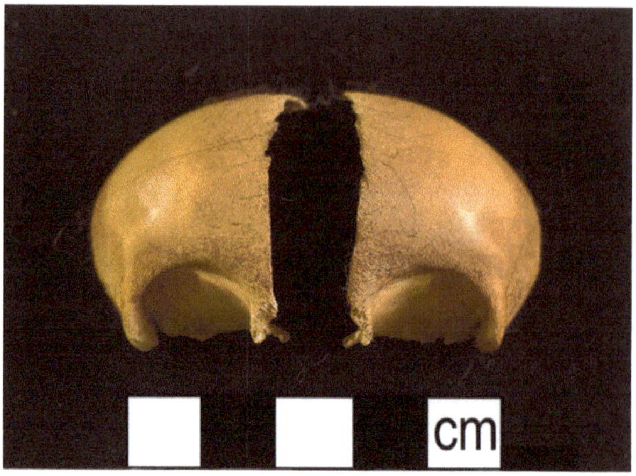

Fig. 1.2 Unfused frontal bones from Gremlin's infant

The skeleton is complete for a chimpanzee neonate, except for there being no ribs associated with this skeleton. Secondary ossification centers and cranial elements are all unfused (Fig. 1.2), and no teeth are erupted. Like Melissa's infant, described above, the periosteal layer of cortical bone was still in the process of formation on his cranial elements. There are no signs of trauma or pathology on his skeleton.

Gaia's Infant (2008) (Male: Days Old?)

Gaia, aged 15 years in 2008, gave birth to twin sons in July of that year. Both infants were seized a few days after their birth by Melissa's daughter Gremlin, their maternal grandmother, who has previously been observed to take and (attempt to) care for Gaia's infant (Godot in 2006). Both of the infants born in 2008 died. The smaller of the infants always looked feeble, was a poor clinger, and may have died within a few days or even hours after birth as the soft tissues were already heavily autolyzed at the time of necropsy. Gremlin eventually dropped the body while climbing a tree, and Gaia re-claimed her already dead infant and carried him around for a few more days. Eventually she abandoned the body and it was recovered by field observers. The brother's body was, unfortunately, never recovered (GSRC).

The infant's skeleton is at the developmental stage expected for a chimpanzee neonate. The frontal bones and mandible are unfused. The post-cranial skeletal elements are also unfused, with no development of secondary ossification centers. No teeth are erupted. The crowns of 13 deciduous teeth in the process of forming were recovered, as were 19 metacarpals /metatarsals, 19 carpals/tarsals, and 43 phalanges (all types).

No significant soft tissue lesions were noted (Karen Terio unpublished data); though a few skeletal features are of potential significance.

Like Melissa's infant in the previous case, this neonate exhibits incomplete formation of the periosteal surface of many cranial bones (Fig. 1.3) as well as the mandible (Fig. 1.4). Because this infant is a twin, researchers at Gombe hypothesize

Fig. 1.3 Palatal view of the unfused maxillae from Gaia's infant

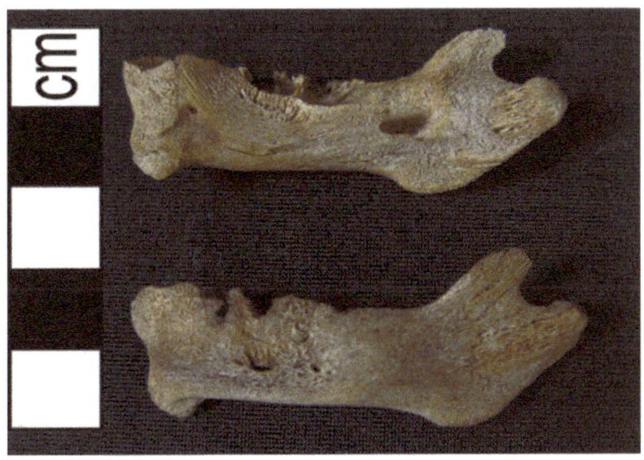

Fig. 1.4 Unfused mandible from Gaia's infant. Top: medial view of right side, bottom: lateral view of left side

that Gaia may have delivered pre-term. More of the skeletal elements of this infant have exposed trabecular bone than do Melissa's infant or Patti's infant (described below); the cranial vault in particular exhibits porosity in the outer table. This infant is hypothesized to be less skeletally mature than Melissa's infant or Patti's infant, though their post-partum ages at death are likely similar.

Patti's Infant (1978) (Male: 1 Week)
The death of this male infant is attributed to lack of maternal care (Goodall 1986). This was Patti's first recorded birth; when she was estimated to be 18 years old (Ibid.). The infants of primiparous females are at higher risk of death compared to subsequent pregnancies (Williams et al. 2008; Goodall 1986). The skeleton is complete for a chimpanzee neonate, and includes 26 total vertebral bodies (likely including sacral) and 33 phalanges. All 20 metatarsals and metacarpals are present. The frontal bones (Fig. 1.5) and mandible are unfused. The post-cranial skeletal elements are also unfused, with no development of secondary ossification centers. No teeth are erupted. Similar to Melissa's infant and Gaia's infant, described above, the maxilla shows incomplete ossification of the periosteal surface of the bone.

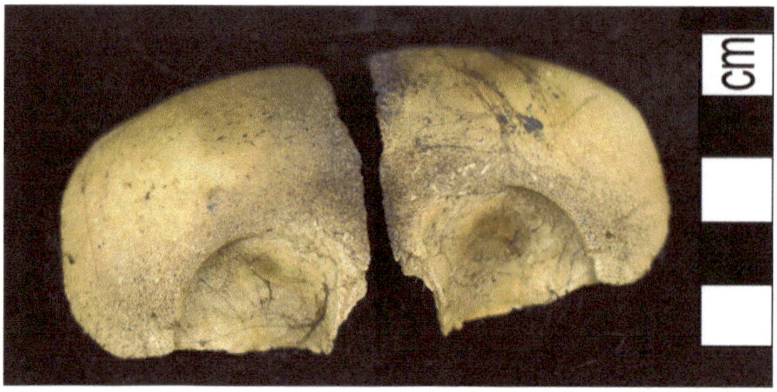

Fig. 1.5 Unfused frontal bones from Patti's infant

Fig. 1.6 Frontal bone from Rejea, exhibiting compression fractures

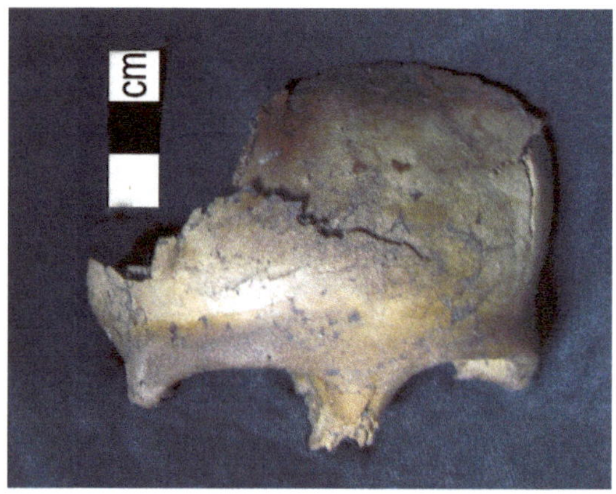

No skeletal traumata or pathologies are evident on Patti's. Patti, who died in 2005, is also part of the Gombe chimpanzee skeletal collection, and is described later in this chapter.

Rejea (Female: 3 Months)
A female infant from the Mitumba community, Rejea was killed and partly consumed by males from the Kasekela community in 1993 (Wilson et al. 2004). Her frontal bone and mandible are fused while the occipital and temporal bones are unfused. Her deciduous incisors and first premolars are in full occlusion.

Rejea is represented by a partial skeleton, due to being partly consumed at the time of death. Neither the scapulae nor fibulae are preserved. The right clavicle, right humerus, and left radius are preserved, but not their contralateral counterparts. Both ulnae, ossa coxae, femora, and tibiae are present, but are fragmentary/broken, as are most of the axial elements. Only two carpals/tarsals and 13 metatarsals/metacarpals are preserved. The hyoid is preserved, though it lacks one greater cornu.

Her skeleton shows damage patterns characteristic of chimpanzee consumption of primates, including compression fractures to the cranial vault (Fig. 1.6), step

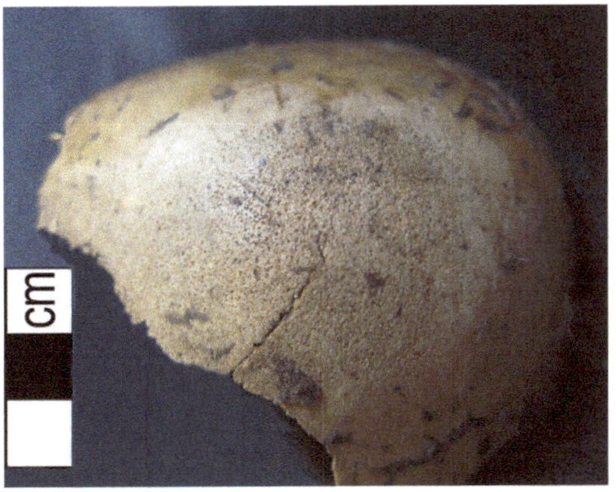

Fig. 1.7 Rejea's parietal bone

fractures of the long bones, frayed long bone ends, and incipient fractures on the ribs (Kirchhoff et al. 2018). Both parietals show increased porosity in a discrete area centered on the parietal bosses (Fig. 1.7), which likely represents incomplete ossification of the outer table due to skeletal immaturity (Scheuer and Black 2004) rather than a pathological condition.

November 1975 Infanticide (Female: 8 Months?)
This unknown female infant was killed by males from the Kasekela community in November 1975 (Goodall 1977). At the time, observers estimated this female infant to be 1.5–2 years old (Ibid.), but comparison of skeletal and dental maturation to a chimpanzee infant of known age (Andromeda) suggests that the infant was younger; the actual age is closer to 8 months (Kirchhoff 2010).

On the day of the attack, a field assistant followed a party of Kasekela chimpanzees ($n = 5$ adult males, two (2) adult females and their four (4) immature offspring) towards the south of their range. The Kasekela chimpanzees encountered an unknown number of strangers. Most of the Kasekela males chased after and attacked the strangers, while a stranger mother with a female infant and a juvenile of unknown sex climbed a large tree, where she remained trapped until the Kasekela males returned and attacked the stranger female. While being attacked, the female dropped her infant, who was seized by one of the males. After 8 min of seeming to battle for possession of the infant, the mother escaped, bleeding heavily. The infant was flailed against branches and rocks, and used in displays. After this, another male picked up the still-living infant and groomed her. The infant was carried by three different Kasekela chimpanzees for several hours, and then abandoned. A research assistant collected the infant, who died later that evening (Goodall 1977). The infant was not consumed, though there was opportunity to do so (Kirchhoff et al. 2018).

The skeleton is nearly complete, lacking only some axial elements and hand/foot bones (lack of the latter is likely due to incomplete ossification as well as preservation/recovery bias). There are three tarsals/carpals association with this skeleton, 25 total phalanges, 23 vertebrae, and 25 ribs. None of the secondary ossification centers are present. The mandible and frontal bones are completely fused, and the deciduous incisors and first premolars were all in occlusion at the time of death.

The right scapula, ilium, radius, both ulnae, three left ribs, and one proximal manual phalanx all show possible incipient fractures. The ribs in particular (Fig. 1.8) show a damage pattern common in chimpanzee infanticide victims (Kirchhoff et al. 2018).

October 1975 Infanticide (Male: 9 Months?)

This unidentified male infant was killed on 30 September 1975 (GSRC data), though the publication which describes the incident (Goodall 1977) indicates that it occurred in October of that year. Shortly after hearing the sounds of chimpanzees fighting, observers encountered a group of chimpanzees from the Kasekela community, handling the still-bleeding infant. Based on the infant's injuries and the types of calls heard (fighting sounds rather than predator alarm calls), Bambanganya (one of the field assistants) logically concluded that the infant had been recently killed by the adult chimpanzees. The carcass was used in displays and partially consumed (Goodall 1977).

This is a partial skeleton. There are no radii or ulnae associated with this individual, and only one tibia and one fibula. They are so fragmentary that side is difficult to determine. Both ossa coxae and femora are present, but all are fragmentary. The mandible also exhibits perimortem breaks. There is only one carpal/tarsal preserved, and no other hand/foot bones. There are 25 vertebral fragments and 37 rib fragments associated with this skeleton.

Skeletal damage is consistent with consumption patterns associated with chimpanzee infanticide victims, including step fractures and crenulation/fraying of long bone ends, puncture wounds to the cranial vault, and incipient fractures on ribs (Kirchhoff et al. 2018). During consumption of chimpanzee infanticide victims, the long bones are often chewed on the ends, causing the fraying of the ends. Flailing and display behavior likely causes the incipient fractures observed on chimpanzee

Fig. 1.8 Incipient fractures on the ribs of an infanticide victim

infanticide victims' axial skeletons (especially ribs). The puncture wounds in this infant's frontal bone (Fig. 1.9) and left parietal are consistent with bite(s) to the head, which is a common method of killing chimpanzee infanticide victims (Arcadi and Wrangham 1999; Kirchhoff et al. 2018).

Age at death is approximately 9 months based on skeletal size and dental development (similar to Andromeda, a known-age infant). The deciduous incisors and first premolars are all erupted, while the frontal bone and mandible are both fused. Long bones are estimated to all be unfused, though none of the preserved long bones are complete.

Andromeda (Female: 9 Months)

Andromeda is another female infanticide victim from the Mitumba community. Her death was not directly observed, though males from Kasekela were followed into the Mitumba community's territory and field observers heard sounds consistent with an aggressive intercommunity encounter. Andromeda was not consumed, but was re-claimed by Mitumba males subsequent to (or not long before) her death. Her body was observed to be used in displays, carried, and groomed before it was eventually abandoned and recovered by researchers (Kirchhoff et al. 2018).

Andromeda's skeleton is mostly complete, though it displays both peri- and postmortem damage. Her right humerus, radius, ulna, and os coxa are fragmented, as are both femora, both fibulae, and both tibiae. There are 14 metapodials, 23 vertebral fragments, and 33 total phalanges associated with her skeleton. Her deciduous incisors and premolars are all in full occlusion, and the deciduous canines are less than half erupted.

Even though she was not at all consumed, Andromeda still shows skeletal damage similar to that of Rejea, including step fractured long bones (Fig. 1.10), incipient rib fractures, and compression fractures and punctures to the cranial vault. Because she was not consumed, Andromeda displays much lower levels of bone fragmentation than Rejea, and better bone survivorship (i.e., her skeleton is more

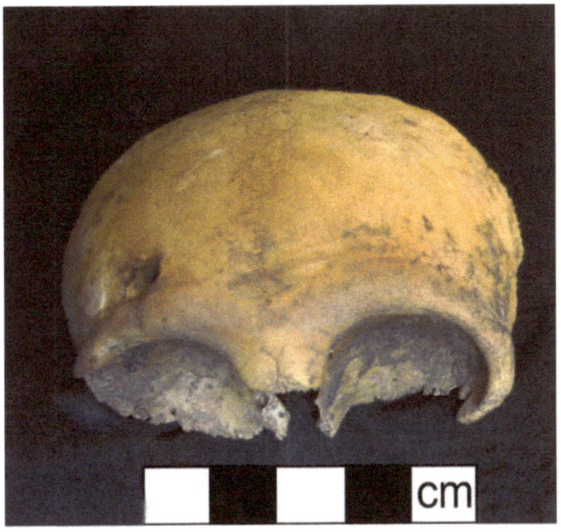

Fig. 1.9 Perimortem puncture wound evident on the right frontal of an infanticide victim

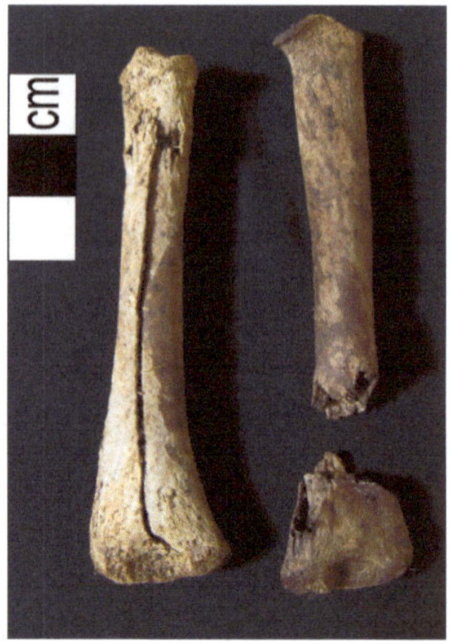

Fig. 1.10 Perimortem damage to Andromeda's humeri

complete than the skeleton of partially consumed Rejea) (Kirchhoff et al. 2018). The surfaces of the bones from Andromeda's skeleton are flaky and bleached in appearance; all elements are very friable. This is likely due to a longer-than-desirable burial period before her skeleton was cleaned.

Gyre (Male: 9.5 Months)

Gyre was one of the twins Melissa delivered in October of 1977, when she was estimated to be 27 years old (Goodall 1986; GRSC; Zihlman et al. 1990). This was her seventh known pregnancy. Gyre appeared to be the smaller, weaker of the twins starting from birth, and he died of a respiratory illness in August 1978 (Goodall 1986). Respiratory illness is the most common cause of death for chimpanzees at Gombe (Williams et al. 2008).

Researches hypothesized that Melissa was unable to produce sufficient milk for her twins, given their apparently slow rate of growth and small body size. It is possible that her infection with poliomyelitis during the 1966 outbreak at Gombe affected her milk production abilities (Zihlman et al. 1990) (or, less directly, her own ability to forage).

Gyre's skeleton is mostly complete; some axial elements have not been preserved (only six vertebrae are associated with this skeleton), and only the ilia represent the pelvis. Epiphyses are all unfused and secondary ossification centers have not developed. The frontal bones are fused, but the metopic suture is still clearly visible (Fig. 1.11).

The deciduous central incisors are fully erupted, while the lateral incisors are partly erupted (maxillary and mandibular). The maxillary first deciduous premolars are in full occlusion while the mandibular counterparts were lost postmortem. Alveolar morphology suggests these teeth were at least partly erupted at the time of

Fig. 1.11 Metopic suture on Gyre's frontal bone

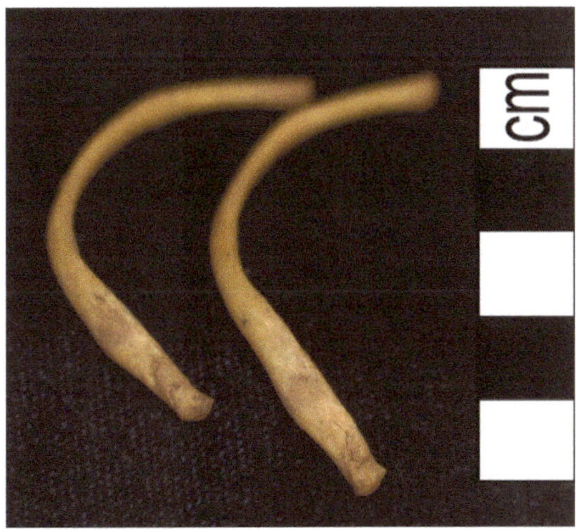

Fig. 1.12 Possible healed fractures near the vertebral ends of two of Gyre's ribs

death. The deciduous canines and second premolars (maxillary and mandibular) are visible in their crypts but unerupted, as are the first permanent molars (all four).

Gyre shows evidence of healed fractures on two of his right ribs. A small, well-healed callus near the vertebral end of both ribs and the altered contour associated with the callus are consistent with healed blunt force trauma (Fig. 1.12). Common causes of this kind of trauma in chimpanzees include falls and inter-individual aggression (Jurmain 1989, 1997).

Fred (Male: 10.8 Months)

Fifi's infant Fred died during the mange outbreak at Gombe in 1997 (Williams et al. 2008). Fifi is one of Flo's offspring; Flo is also part of the skeletal collection and her case is described later in this chapter.

Fred was Fifi's seventh pregnancy, born when she was estimated to be 38 years old. Fifi had two pregnancies following Fred: Flirt, a female born in 1998, who is still alive, and Furaha, another female born in 2002 and presumed to have been killed by intercommunity violence in 2004 (GSRC data). Genetic testing revealed that Frodo was Fred's father. Since Frodo is also Fred's half-brother (Constable et al. 2001), lack of genetic diversity may have contributed to Fred's cause of death, though certainly the possibility of decreased milk production while his mother was ill during the mange outbreak could be a factor in addition to Fred's illness.

Fred's skeleton is nearly complete, lacking only a few bones. There are 16 metapodials, 24 vertebrae, 23 ribs, and 45 phalanges associated with the skeleton, as well as the body of the hyoid and one greater cornu. Secondary ossification centers are not yet developed. His deciduous incisors and premolars are all erupted, though one of the maxillary second premolars is not yet in full occlusion. The deciduous canines are less than half erupted. Fred also exhibits a metopic suture in the glabellar region (it does not extend superiorly) (Fig. 1.13).

Pathologic lesions on Fred's skeleton involve abnormal bone formation. This includes periosteal reactive bone on four metapodials and three manual intermediate phalanges, which, as a general sign of inflammation, is consistent with his cause of death. His temporal glenoid fossae exhibit abnormal porosity, though they do not meet the criteria for temporomandibular disease in humans (Rando and Waldron 2012).

Groucho (Male: 1.28 Years)

Groucho was born in July 1985; the seventh of Melissa's recorded pregnancies, he was born when Melissa is estimated to have been 35 years old. Groucho died of "wasting disease" in October of 1986, 2 weeks before his mother died of a similar illness (GSRC). Melissa's case is described later in this chapter. Wasting disease is defined as "noticeable weight loss and weakness before death" (Williams et al. 2008,

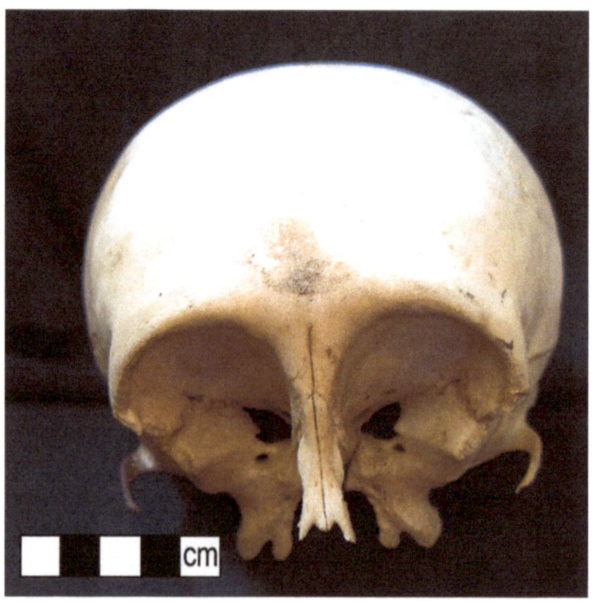

Fig. 1.13 Fred's metopic suture is not fully fused. The maxillae are also unfused (not pictured)

p. 770), and is the second-most common cause of illness-related death among Kasekela chimpanzees (respiratory illness is the most common) (Ibid.).

The skeleton is mostly complete, with secondary ossification centers developed, but all unfused. The skeleton preserves 16 total metacarpals/metatarsals, 25 phalanges, four tarsals, 13 vertebrae, and 26 paired ribs. The deciduous incisors and premolars are all fully erupted, while the deciduous canines are beginning to erupt. The permanent incisors and first molars are visible in their crypts but unerupted (Fig. 1.14).

There are no signs of trauma or pathology on the skeleton, but of note is that the mandibular symphysis is unfused (Fig. 1.15). This is counted as a "congenital anomaly" since much younger infants from Gombe have fused mandibles (e.g., Rejea, who was 3 months old at death).

Plato (Male: 2.61 Years)
Plato was born in April 1973. He was Pallas's first recorded pregnancy, born when she is estimated to have been 15 years old. Plato is represented by the skull and clavicles only; he died of respiratory illness/gastric enteritis in 1973 (Goodall 1986; GSRC). These are the most common causes of illness-related death among the Gombe chimpanzees (Williams et al. 2008). The cranium is fused, with the metopic suture still visible near glabella (Fig. 1.16), but not farther superiorly. The complete

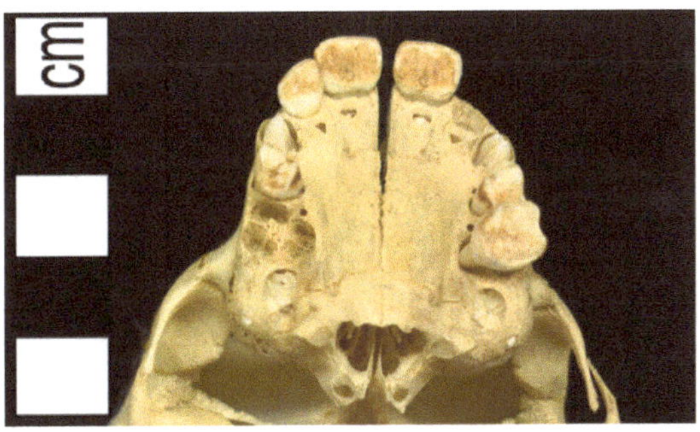

Fig. 1.14 Groucho's maxillary dentition

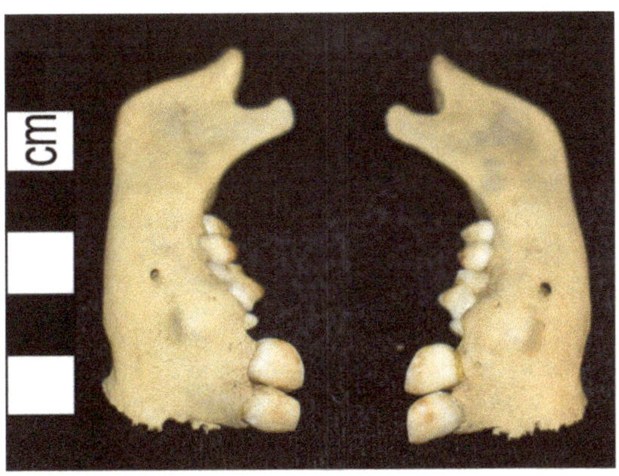

Fig. 1.15 Groucho's unfused mandible

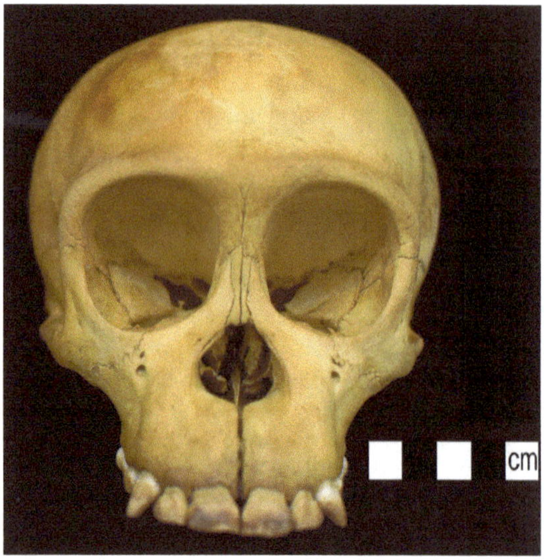

Fig. 1.16 Plato, with metopic suture still partially patent

deciduous dentition is in full occlusion, and the first permanent molars are visible in their crypts. There are no skeletal signs of trauma or pathology, though the mandible exhibits postmortem fractures.

Pallas is also part of the Gombe chimpanzee skeletal collection; her case is described later in this chapter.

Getty (Male: 3.89 Years)

Getty is the first recorded pregnancy for Gremlin (Melissa's daughter); he was born in May 1982, when Gremlin was 12 years old. Getty was killed by poachers in April of 1986. Poaching is less of a problem at Gombe compared to some other long-term primate study sites (e.g., poaching has been continuously combatted at Virunga National Park in Rwanda (Robbins et al. 2011)), though still a troubling phenomenon worth describing and understanding, as a means to prevent it. Tumaini, an adult male described later in this chapter, was also killed by poachers.

Getty's post-cranial skeleton is mostly complete; it lacks the left clavicle, humerus, radius, and ulna. There are 15 total metatarsals/metacarpals, 17 phalanges, and ten tarsals/carpals associated with this skeleton. Some epiphyses are present, including the right proximal humerus, right distal radius, both distal femora, and right proximal fibula. The skull is not part of the skeletal collection; presumably it was retained by the poachers.

Getty exhibits evidence of perimortem (unhealed) fractures to the vertebral ends of both first ribs; the left also exhibits a chop mark on the sternal end. In addition, two left and six right ribs exhibit incipient fractures near the sternal end, and two more left and one more right ribs exhibit longitudinal fractures/splitting. There are step fractures on 13 phalanges and one metacarpal. An incipient fracture or chop mark is evident on one metatarsal. The distal shaft of the left tibia exhibits a chop mark and there is damage to both scapular blades.

In addition, the neural arches of sacral vertebrae 2–5 are open (Fig. 1.17). Two thoracic vertebrae are fused at their spinous processes (Fig. 1.18). The two vertebral bodies are still separate, and all 13 pairs of ribs are present and exhibit normal morphology.

Ebony (Male: 8.2 Years)

An adolescent male from the Mitumba community, Ebony died in 2005. His body was discovered shortly after an incursion by the Kasekela males into the Mitumba community's territory, and it showed soft tissue wounds consistent with chimpanzee attack (M.L. Wilson unpublished data).

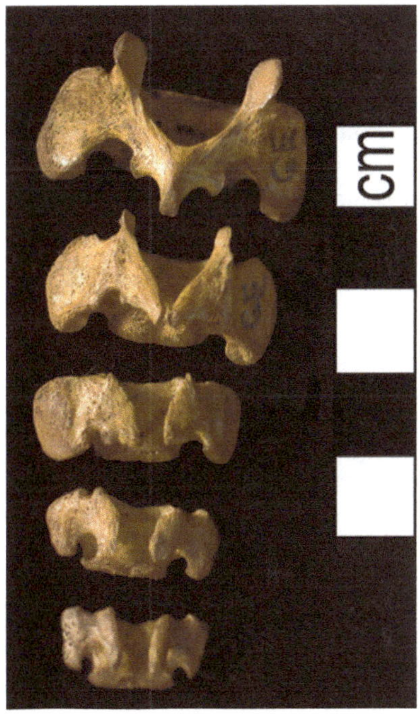

Fig. 1.17 Unfused neural arches in sacral vertebrae 2–5, Getty

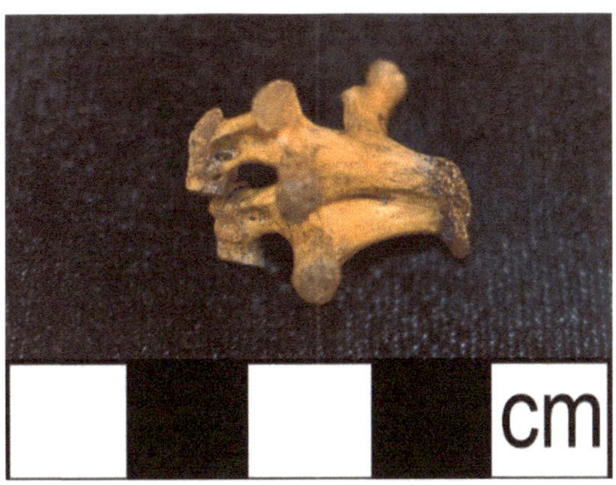

Fig. 1.18 Fusion of the spinous processes of two of Getty's thoracic vertebrae

The skeleton is complete, lacking only the hyoid and a few hand/foot bones. There are 48 phalanges and 30 carpals/tarsals associated with this skeleton. There are well-developed secondary ossification centers, and none of the epiphyses are fused. The inferior pubic ramus is fused to the ischium, and the second through fifth sacral vertebrae have begun to fuse (the first and second remain separate). The deciduous canine is still present, and the third molars are unerupted. The mandibular first premolars are over half erupted, and the rest of the adult dentition is in full occlusion (Fig. 1.19).

As a young individual, Ebony shows few signs of skeletal pathology or trauma. Most notably, his skeleton shows no signs of perimortem trauma. This is particularly striking because soft tissue damage noted shortly after his death was extensive (GSRC).

Two proximal manual phalanges show fractures in the process of healing (antemortem trauma). The femoral (Fig. 1.20) and tibial surfaces of both knees have possible subchondral bone cysts, a lesion whose etiology is likely to follow one of two trajectories, both of which relate to bone development. A subchondral bone cyst may be epiphyseal cartilage that becomes "trapped" in the ossifying epiphysis (Thompson 2007), or may develop as a result of synovial fluid leaking through

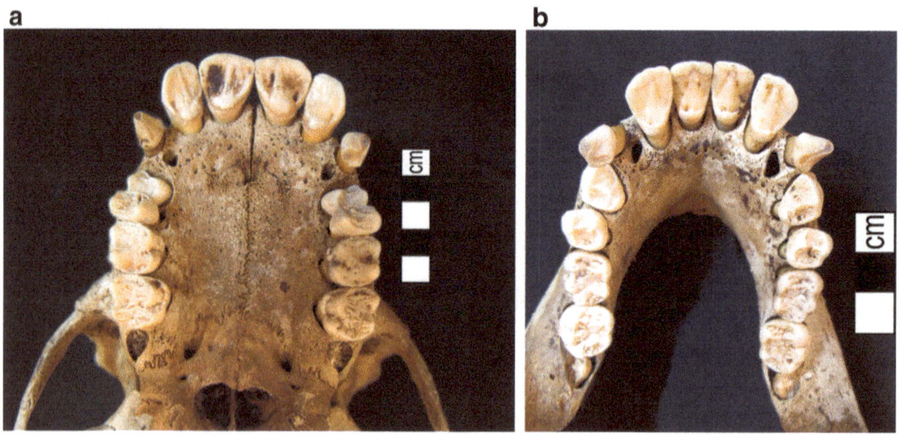

Fig. 1.19 (**a**) Ebony's maxillary dentition. (**b**) Ebony's mandibular dentition

Fig. 1.20 Distal femoral epiphysis from Ebony, exhibiting subchondral bone cysts

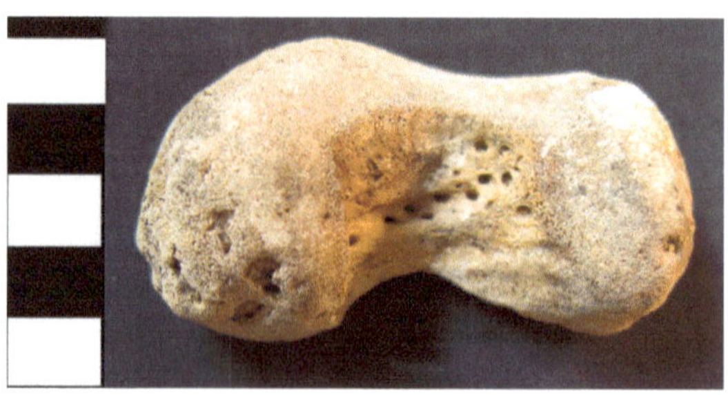

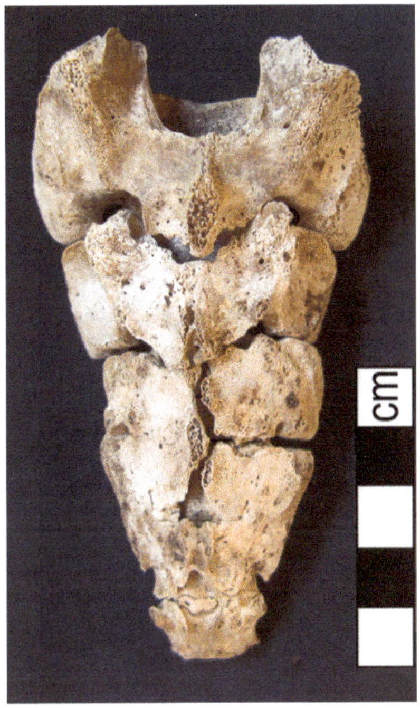

Fig. 1.21 Incomplete neural arches on the second and third vertebrae of Ebony's sacrum

damaged articular cartilage and then becoming trapped in the epiphysis once the bone is completely ossified (Yochum and Rowe 2005). This kind of cyst may also be a sequel to degenerative joint disease in some species (Weisbrode 2007), though this seems unlikely in Ebony's case due to his young age and lack of other evidence of joint degeneration. Laminar deficiency is evident in the second and third sacral vertebrae (Fig. 1.21).

Flint (Male: 8.55 Years)
Flint was born in March 1964 to Flo, when she was estimated to be 45 years old. This was her fourth recorded pregnancy, but it is likely, given her age, she had been pregnant prior to the start of field work at Gombe in 1960. Flint was Flo's second-to-last offspring, and he remained dependent on his mother longer than most young chimpanzees. After his infant brother (Flame) died at 6 months of age, Flint, then aged 4.5 years, continued to sleep with and be carried by his mother, and even nurse until she died 3 years later in 1972, unusual for a juvenile of his age. He died within a month of his mother, and his cause of death is attributed to being orphaned. Flint demonstrated lethargy, loss of appetite, sunken eyes, and gastric trouble in addition to behavior changes such as decreased play and increased nervousness around adult males after his mother died. Comparably aged chimpanzees whose mothers died have recovered and survived well after their mothers; Flint's case is considered exceptional (Goodall 1986). Chimpanzees 4–5 years of age are typically more independent, resulting from a combination of their more individual behaviors as well as the mother's efforts to wean and otherwise encourage independence in their

offspring (often necessitated by the arrival of another infant) (Goodall 1986; Zihlman et al. 1990).

Flint's skeleton is mostly complete, lacking only a few small bones/epiphyses. Secondary ossification centers are present but unfused. Epiphyses likely lost post-mortem include the acromion, capitulum, proximal ulna, proximal fibular, and distal fibula (all bilateral). Associated with Flint's skeleton are four carpals, 12 tarsals, and 38 total phalanges. His permanent incisors are fully erupted, as are the permanent first molars. The permanent second molars are partly erupted, except for the right mandibular, which is in full occlusion. The deciduous canines and premolars are all still in occlusion, except for the right deciduous second premolar, which has been shed. The left is still in situ. Permanent canines and premolars are visible in their crypts.

Flint shows increased porosity of the temporal glenoid fossae (bilateral), as well as mild crowding of his mandibular incisors. There is a destructive lesion near the proximal epiphysis of an intermediate manual phalanx, and the neural arch of his fifth sacral vertebra has abnormal morphology. The skeleton has no other evidence for lesions.

Jackson (Male: 10.4 Years)

Bacterial pneumonia is the suspected cause of death for Jackson, who exhibited symptoms of a respiratory infection and died during a pneumonia outbreak at Gombe in 2000. Respiratory illness is the most common cause of death for Kasekela chimpanzees (Williams et al. 2008). Jackson was the second recorded pregnancy for Jiffy, who is estimated to have been 14 years old when Jackson was born. Jiffy was first observed in the Kasekela community when she was estimated to be nearly 11 years old. Her first recorded infant (unknown sex), born in 1988, disappeared and did not survive the first year of life.

Jackson's skeleton is complete, lacking only a few hand/foot bones. Preserved elements include all long bones and axial elements, 26 total carpals/tarsals, nine metacarpals, nine metatarsals, and 48 total phalanges. The inferior pubic ramus is fused to the ischium, and all other epiphyses are unfused, though secondary ossification centers are well developed. The permanent incisors and first and second molars are in full occlusion. The third molar is not erupted. The deciduous canines are all still in situ, as are the maxillary deciduous premolars. The left first mandibular permanent premolars have fully erupted, and the second is visible in its crypt. The deciduous counterpart may have already been shed prior to death and is not preserved. The right mandibular deciduous premolars are still in full occlusion, though the first permanent premolar is visible in its crypt.

Jackson shows no signs of skeletal trauma, but several pathologies. The distal articulation of the left tibia displays a small (approximately 0.5 mm in diameter) lesion that may have a similar etiology as the proposed subchondral cysts described for Ebony (see above). The distal articulations of both ulnae display a bone-loss pathology of unknown etiology, and the right clavicle exhibits a lesion related to pathological bone loss on its inferior, lateral surface. A manual intermediate phalanx shows abnormal bone formation consistent with periosteal reaction. Given the cause of death, skeletal signs of inflammation are not surprising.

Both tali have extensive lesions on the superior, distal articular aspect. The new bone formed is waxy in appearance, though also very porous (Fig. 1.22). Jackson also displays laminar deficiency in the second sacral vertebra (Fig. 1.23).

Mel (Male: 10.7 Years)

Mel was born in January 1984 to Miff. He was Miff's third pregnancy, born when she is estimated to have been 28 years old. Miff is also part of the Gombe chimpanzee skeletal collection, as is another of her offspring, Michaelmas; their cases are described later in this chapter.

Fig. 1.22 Abnormal bony formation on Jackson's calcaneus. The new bone is porous and waxy-looking

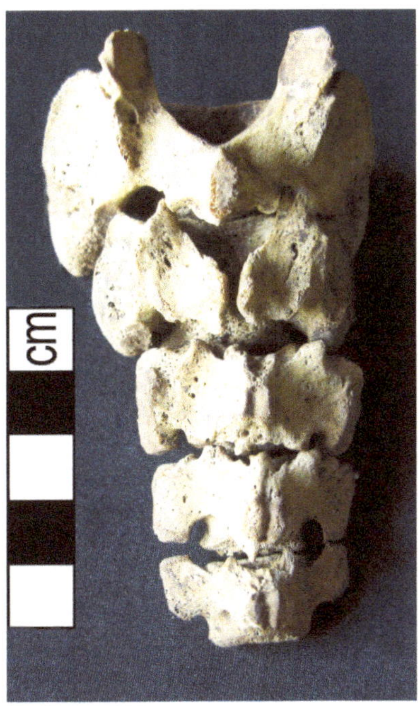

Fig. 1.23 Laminar deficiency in Jackson's second sacral vertebra

Mel's body was discovered in October1994 exhibiting soft tissue injuries consistent with either a leopard predation attempt or chimpanzee attack (Williams et al. 2008). The fact that he was not at all consumed suggests that a leopard attack is less likely (Wilson et al. 2004). The skeleton is complete, lacking only a few small bones. There are 30 carpals/tarsals and 47 phalanges associated with this skeleton, and no coccygeal vertebrae.

The epiphyses are well developed, but unfused. The inferior pubic ramus has fused to the ischium, and the acetabular ossification center between the three bones of the os coxa is in the early stages of unification. Most of the adult dentition is in full occlusion, exceptions being the unerupted permanent third molars and the adult canines. The maxillary deciduous canines are still in situ, while the mandibular deciduous canines were shed antemortem. The left canine has not started erupting, nor is there a patent alveolus. A bulge in the mandibular corpus seems to indicate that the canine is, in fact, developing, but a radiograph would be necessary to confirm. The mandibular incisors are crowded (Fig. 1.24), suggesting a possible discordance between mandibular versus dental development (see Zihlman et al. 1990 for commentary related to Flint).

Two phalanges (one manual and one pedal) show evidence for well-healed fractures. His maxilla shows an area of reactive bone near prosthion (Fig. 1.25) consistent with an area of localized infection. The anterior maxillary dentition, which was recovered when the skeleton was exhumed, does not, however, show signs of breakage or other pathology. The alveolar bone near the left maxillary first molar shows evidence for periodontal disease, as the bone on the palatal side shows a smooth-edged erosion that widens the alveolus (Fig. 1.26).

Fig. 1.24 Crowding of Mel's mandibular incisors

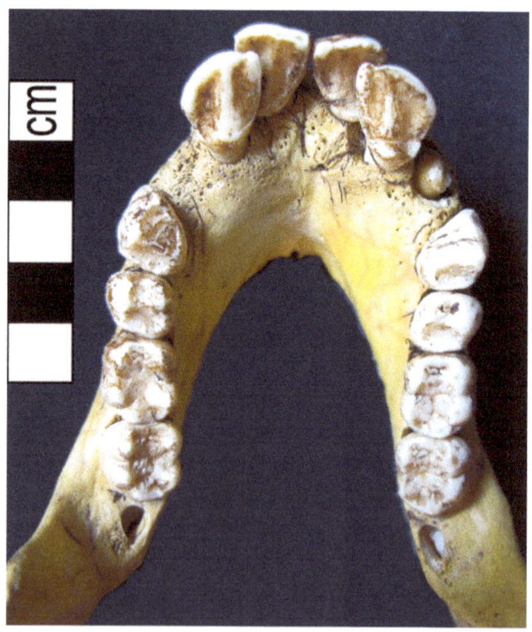

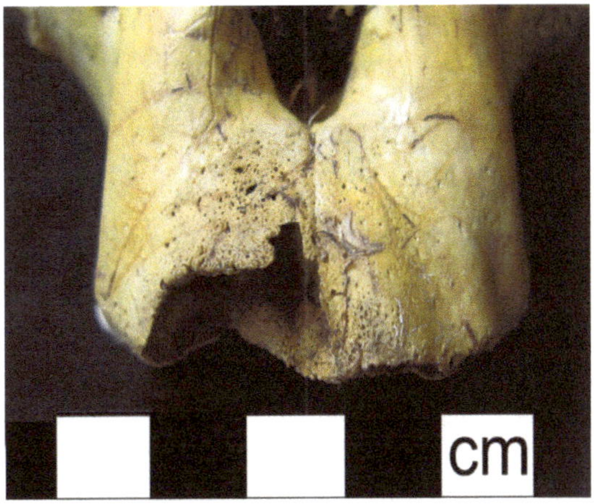

Fig. 1.25 Reactive bone on Mel's maxilla

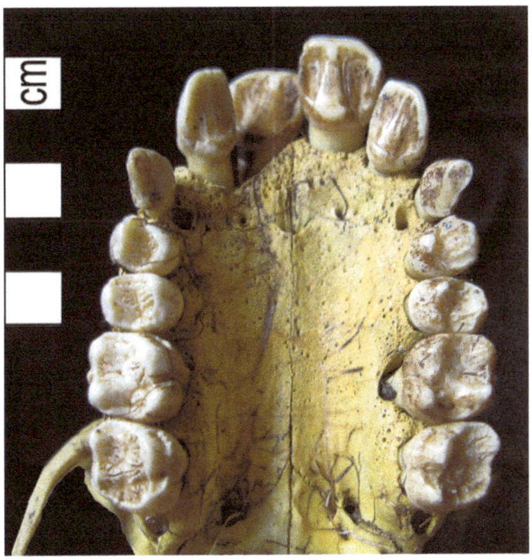

Fig. 1.26 Pocketing on the palatal aspect of the alveolus of Mel's upper left M1

Three carpals, both distal femora, and both articular surfaces of the right tibia display round lesions approximately 2–3 mm in diameter that may be evidence for subchondral cysts (see *Ebony* for more detail). The right fifth metatarsal shows proliferative bone consistent with periosteal reaction.

Mel exhibits laminar deficiency in the third, fourth, and fifth sacral vertebrae. The sacral vertebrae are all unfused. There are no skeletal signs of perimortem trauma.

Sugar (Female: 10.9 Years)

Sugar (alternate spelling: *Shuga*) was born to Caramel in July 1976, when Caramel is estimated to have been 20 years old. Sugar was Caramel's third pregnancy.

Sugar died of a respiratory infection in May 1987, during an outbreak at Gombe (Williams et al. 2008). Her nearly complete skeleton is well preserved and shows few signs of pathology or trauma. The left maxilla, hyoid, and distal epiphysis of the left femur are not preserved. There are 23 carpals/tarsals and 29 phalanges associated with this skeleton. The maxilla and other facial bones have not fused with the neurocranium, which is notable as both Jackson and Mel (of similar ages) have fused crania. The epiphyses are mostly unfused and secondary ossification centers are well formed. The inferior pubic ramus has fused to the ischium, and the distal epiphysis of the left humerus is in the early stages of fusion. The adult canines and third molars are the only unerupted permanent teeth. The maxillary deciduous canines are still in occlusion, but the permanent mandibular canines are beginning to erupt. The mandibular incisors are very slightly crowded.

A lesion on the distal articulation of a pedal proximal phalanx (4.3 × 0.4 mm) shows abnormal bone resorption; its etiology is unknown. It is not morphologically consistent with a subchondral bone cyst. Interesting congenital variations include patent olecranon foramina on both humeri and asymmetrical transverse foramina on the first cervical vertebra. Sugar displays laminar deficiency in the third and fourth sacral vertebrae (Fig. 1.27).

Galahad (Male: 11.8 Years)

Galahad is Melissa's grandson, and Gremlin's third pregnancy. He was born in April 1988 when Gremlin was 18 years old (Goodall 1986). He is suspected to have died of pneumonia in February 2000, during an outbreak of respiratory illness at Gombe (Williams et al. 2008). Besides the adult canines, which are less than half erupted,

Fig. 1.27 Laminar deficiency in Sugar's third and fourth sacral vertebrae

only the maxillary incisors and the mandibular lateral incisors are preserved, and eruption of the other teeth is estimated from the alveoli. The third molars were at least half erupted at the time of death. The rest of the adult dentition (other than the canines) was in full occlusion at death. The secondary ossification centers are well formed but largely unfused. The distal humerus is less than half fused, as is the right medial epicondyle. The inferior pubic ramus is fused to the ischium. The skeleton lacks only a few hand/foot bones, the coccyx, and one greater cornu of the hyoid. There are 23 carpals/tarsals, nine metacarpals, nine metatarsals, and 41 phalanges associated with this skeleton. Galahad's complete vertebral column includes 12 thoracic vertebrae (with 24 paired ribs) rather than the more usual 13, and four lumbar vertebrae.

Subchondral bone cysts affect several articular surfaces in the hands as well as the left talus. The articular surface of the left patella is abnormally porous and has a shiny appearance. Its appearance is not consistent with the eburnation characteristic of degenerative joint disease, but is indicative of a proliferative lesion. The vertebral end of a right rib shows an unusual articular morphology (Fig. 1.28). The head and tubercle are both fan shaped, and the articular surfaces are irregular. It is possible this is a dislocation deformity subsequent to trauma, though none of the thoracic vertebrae exhibit a corresponding pathology.

A depression on the frontal bone near bregma, measuring approximately 13 × 3 mm, may be evidence of well-healed cranial trauma (Fig. 1.29) (Novak and Hatch 2009). The sixth cervical vertebra is congenitally missing the transverse foramina (Fig. 1.30). In addition, laminar deficiency in the second, third, fourth, and fifth sacral vertebrae is present (Fig. 1.31).

Michaelmas (Male: 12.99 Years)

Michaelmas was the second infant born to Miff, when she is estimated to have been 17 years old (Goodall 1986; GSRC). He was born in October 1973, and died of "wasting disease," or gastrointestinal distress in September 1986 (GSRC; Williams et al. 2008). His skeleton is complete, lacking only the hyoid and a few hand/foot

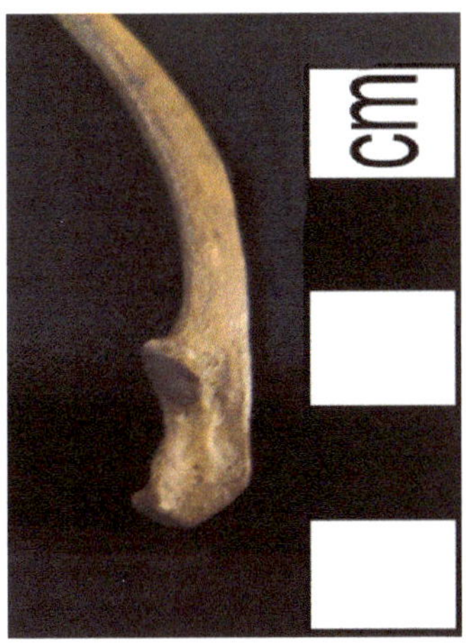

Fig. 1.28 Abnormal morphology of the vertebral end of one of Galahad's ribs

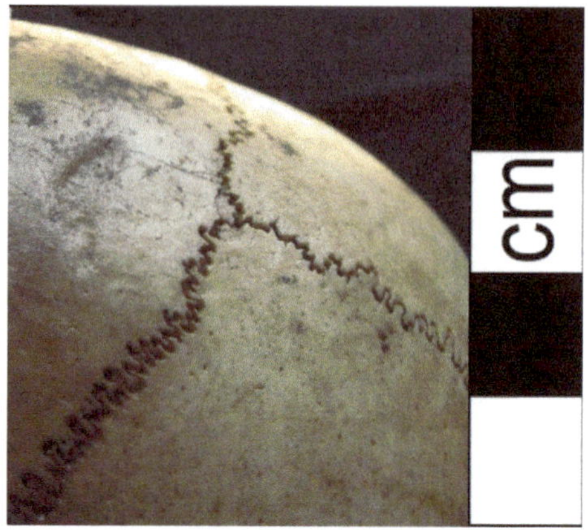

Fig. 1.29 Possible depression fracture near bregma on Galahad's cranium

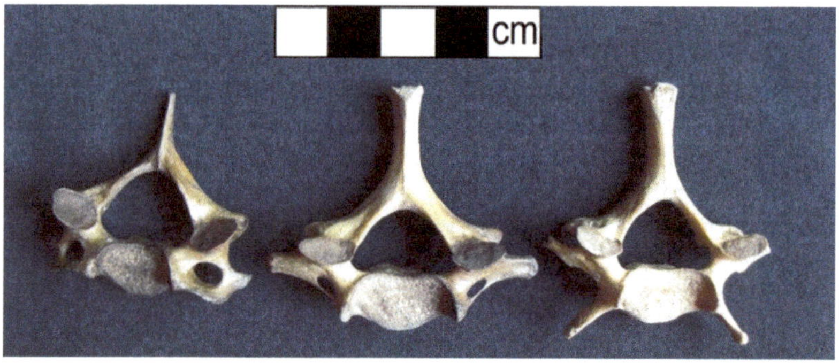

Fig. 1.30 Cervical vertebrae 4, 5, and 6 (left to right) from Galahad. Not absence of transverse foramina on C6

bones. Twenty-nine (29) carpals/tarsals are associated with this skeleton, as well as nine metacarpals, ten metatarsals, and 77 total phalanges.

Michaelmas shows several interesting skeletal traumata/pathologies. Two phalanges exhibit evidence of healed fractures, and his upper right first molar was lost antemortem. Given his age at death, the loss might have occurred subsequent to trauma. The most dramatic lesion for Michaelmas is related to an injury to his left lower limb in 1977 when he was 4 years old, sustained during an attack on his mother, Miff. Michaelmas didn't use the limb for 10 months, and subsequently showed less rapid growth than comparably aged male chimpanzees (Goodall 1986). The skeleton shows evidence for dislocation of the left hip, particularly in the form of extensive remodeling of the left acetabulum (Jurmain 1989). In addition, the right femur is approximately 10 mm shorter than the left.

MacDee (Male: 13.45 Years)

MacDee is estimated to have been born in July 1953, when his mother, Jessica, would have been 13 years old. He was Jessica's first recorded pregnancy. MacDee

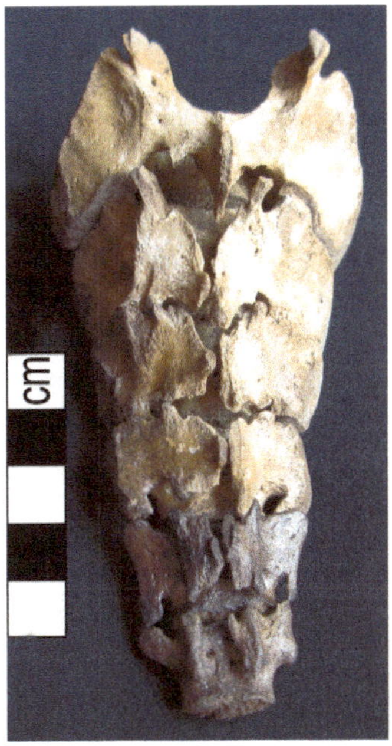

Fig. 1.31 Laminar deficiency in Galahad's sacrum

died in December 1966 (GSRC); a victim of the polio epidemic, he was shot in the head for humanitarian reasons after losing the ability to move both arms and hands (Goodall 1986). His skull exhibits unhealed trauma consistent with gunshot wounds, with one entry wound on his left frontal measuring approximately 8 mm in diameter (patent portion, total extent of damage nearly 21 mm) and the corresponding exit wound in the left occipital measuring approximately 32 mm. A second entry wound on the right ascending ramus of the mandible measures approximately 10 mm, and the left gonial angle exhibits an area of damage measuring approximately 42 mm.

MacDee's skeleton is complete, lacking only a few bones from the extremities: 25 carpals/tarsals are associated with this skeleton, along with ten metacarpals, nine metatarsals, 39 total phalanges, and two coccygeal vertebrae. All other vertebrae (seven cervical, 13 thoracic, four lumbar, six sacral) and 13 pairs of ribs are present.

In addition to perimortem gunshot trauma, MacDee's skeleton exhibits evidence for a few other pathologies of interest. The right calcaneus (inferior articulation) and right trochlea both exhibit small chondroid cysts (bony evidence thereof), and the left pubis has an osteoma approximately 4 mm in diameter. The left humerus has a patent olecranon foramen (3 mm). MacDee exhibits reactive bone consistent with periosteal reaction on one metacarpal, two manual phalanges, both posterior tibiae, the right frontal bone, and six left ribs.

There is also one instance of antemortem trauma: a partly healed probable bite wound on the maxilla measuring 5 mm in diameter (Jurmain 1989). The lesion is not associated with any dental traumata or pathologies. Inter-individual aggression is the second-most common cause of death among male chimpanzees at Gombe

(Williams et al. 2008), and facial trauma is commonly sustained in male agonistic encounters (Novak and Hatch 2009; Jurmain and Kilgore 1998).

Sherehe (Female: 15.78 Years)
In January 1991, Sherehe was born to Sandi, then aged 18 years. This was Sandi's first recorded pregnancy. In November 2006, Sherehe was found with extremely limited mobility and she died soon after. Her cause of death was not immediately known, and disease or a fall from height were both considered possible causes of death. Soft tissue pathological analysis was not possible in this case due to the circumstances of death, though gross examination revealed a ruptured gall bladder (GSRC; Terio et al. 2011). The skeleton does not exhibit any evidence of perimortem trauma, though there are some other bony lesions of potential interest.

The skeleton lacks only some phalanges (48 total were recovered). Sherehe has 14 thoracic vertebrae with 28 paired ribs, and four lumbar vertebrae. Epiphyses are well formed and many are completely fused. The acromion, medial scapular margin, medial clavicle, and iliac crest are all unfused. The proximal humerus is less than half fused. The distal radius, distal femur, proximal tibia, proximal fibula, and sacrum are at least half fused. Fully fused elements where a clear epiphyseal line is still visible include the femoral head and distal tibia. The other epiphyses are fully fused with no visible epiphyseal line. The complete, adult dentition is in full occlusion.

Sherehe exhibits alveolar pocketing consistent with periodontal disease on the buccal aspect of the alveoli for the left mandibular third molar and the right mandibular second molar (Fig. 1.32). Possible subchondral bone cysts were observed on the trochlea of the left humerus (Fig. 1.33) as well as the proximal articulations of two metapodials, the distal pedal phalanx of the right first digit, and the right first cuneiform. Many of the bony joint surfaces exhibit increased porosity, but color changes and the friability of the bone indicate that this is likely to be diagenic change rather than joint disease. This assessment is consistent with Sherehe's young age at death. Two intermediate pedal phalanges exhibit lipping of the proximal articulation that might be consistent with the early stages of joint disease. Laminar deficiency in the second through fifth sacral vertebrae is present (Fig. 1.34).

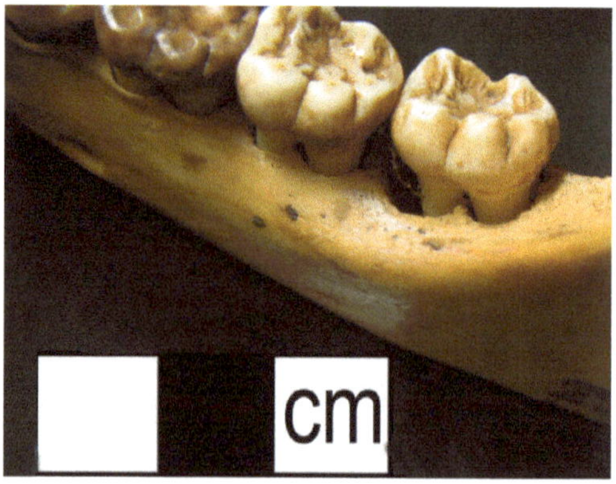

Fig. 1.32 Pocketing of the alveolus for Sherehe's lower left M3, indicative of periodontal disease

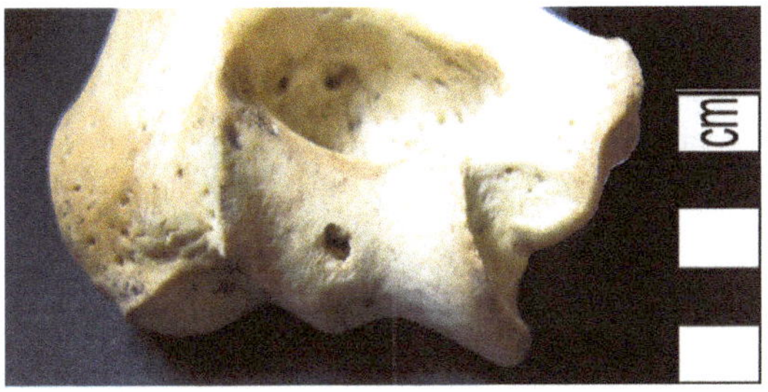

Fig. 1.33 Subchondral cyst on Sherehe's distal humerus

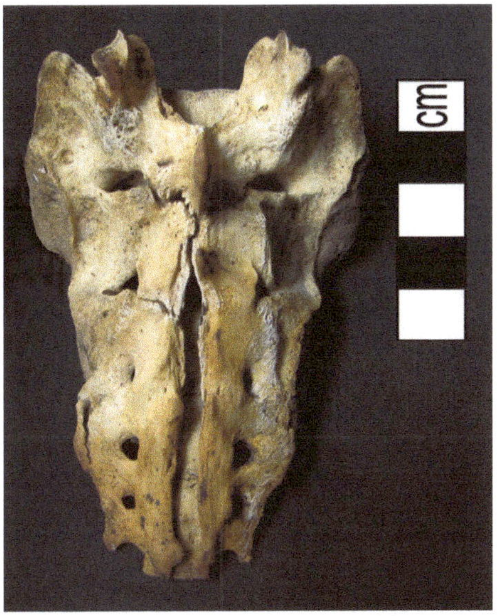

Fig. 1.34 Laminar deficiency in Sherehe's sacrum

Three additional elements were recovered with Sherehe's skeleton. Sherehe was interred in a permeable bag, so it is likely these elements were indeed associated with her skeleton in some way. Two sharp-edged pieces of bone with irregular surfaces (Fig. 1.35) have morphology consistent with ossified portions of costal cartilage. At under 16 years old, ossified costal cartilage would be unusual for Sherehe. An unidentified broken fragment somewhat resembling a tooth was also recovered (Fig. 1.36).

Gilka (Female: 18.9 Years)

Gilka was first observed with her mother, Olly, in 1963, when Gilka is estimated to have been approximately 3 years old. She is Olly's first recorded pregnancy, though given that Olly is estimated to have been 21 when Gilka was born (GSRC), Olly may have been pregnant previously. During her life, Gilka suffered from many ailments, including polio in 1966 and a fungal infection that caused her face to swell beginning in 1968. She had at least two infants stolen from her and eaten by Passion and Pom in 1974 and 1976. Gilka never fully recovered from this second attack, and

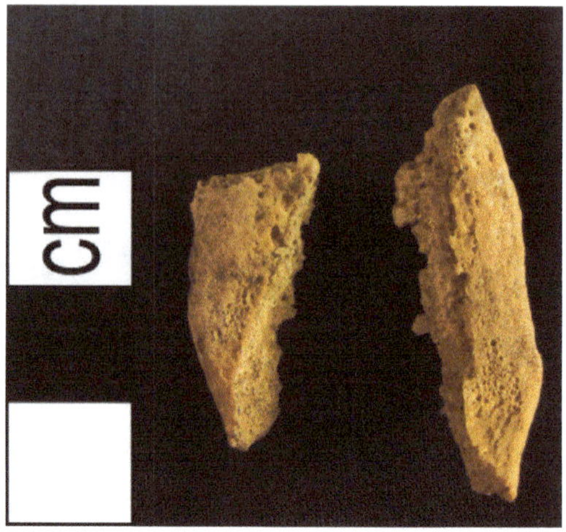

Fig. 1.35 Unidentified osseous fragments associated with Sherehe's skeleton

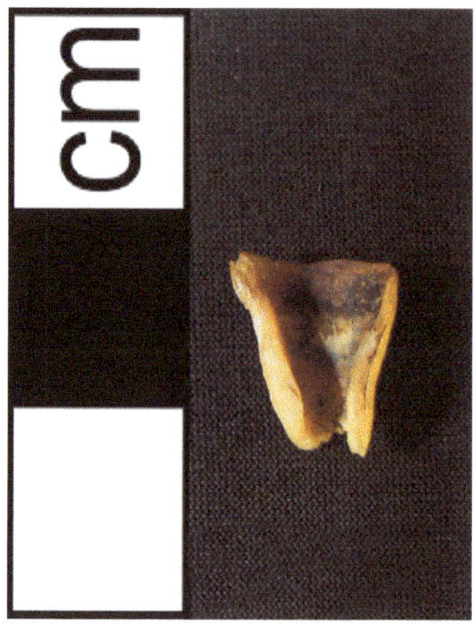

Fig. 1.36 Tooth fragment associated with Sherehe's skeleton

died in May 1979 (Goodall 1986). Her death is attributed to gastrointestinal distress (Williams et al. 2008).

Gilka's skeleton is mostly complete, with 19 carpals/tarsals preserved, seven metatarsals, and 30 total phalanges. Only the body of the hyoid is preserved, and there are no coccygeal vertebrae associated with her skeleton. There are 12 thoracic vertebrae associated with her skeleton, but 26 pairs of ribs. All other skeletal elements are preserved and have well-formed epiphyses, most of which are fully fused. The acromion process, medial margin of the scapula, and medial clavicle are still unfused. The iliac crest is more than half fused, and epiphyseal lines are still evident on the proximal humerus, proximal radius, femoral head, and sacrum. The complete adult dentition is in full occlusion, though the teeth are more worn than is usual for a chimpanzee her age (Kilgore 1989).

Jurmain (1989) has described likely healed fractures to two of Gilka's metatarsals, and erosional lesions on five of her phalanges attributable to a bacterial infection. For at least 2 years before her death, open sores were visible on Gilka's hands and it is unclear whether this infection is of similar etiology as the fungal infection that affected her face (Goodall 1986; Ibid.).

Gilka exhibits joint surface eburnation on the proximal and distal aspects of all five left metacarpals, the left trapezoid, right capitate, one distal and two proximal manual phalanges, the proximal and distal ends of the right humerus, the proximal end of the left humerus, the proximal and distal ends of the right radius, right ulna, left radius, and both femora. Given her young age, it seems likely that these joint pathologies resulted from over-use subsequent to her bout with polio (which primarily affected her right forelimb) and to the deforming infection(s) that affected her hand (Jurmain 1989).

Her right ulna and humerus are noticeably of a decreased diameter compared to the left, a result of her bout with polio (Jurmain 1989; Morbeck et al. 1991; Morbeck 1987). It is possible that her maxillary incisors were broken; they exhibit a higher degree of wear than the mandibular incisors, and the pulp cavity of both maxillary central incisors is exposed. A corresponding periapical abscess drainage site is evident on her maxilla, associated with the right central incisor (Fig. 1.37) (see also Kilgore 1989), and small patch of healed reactive bone on the left side may correspond to a periapical infection of the left central incisor that resolved prior to her time of death. It is possible Gilka used her dentition/mouth more often to manipulate objects as a result of partial paralysis of her right forelimb. Her right humerus exhibits a patent olecranon foramen, and the left navicular has only the articulations for the cuneiforms, lacking the tubercle congenitally.

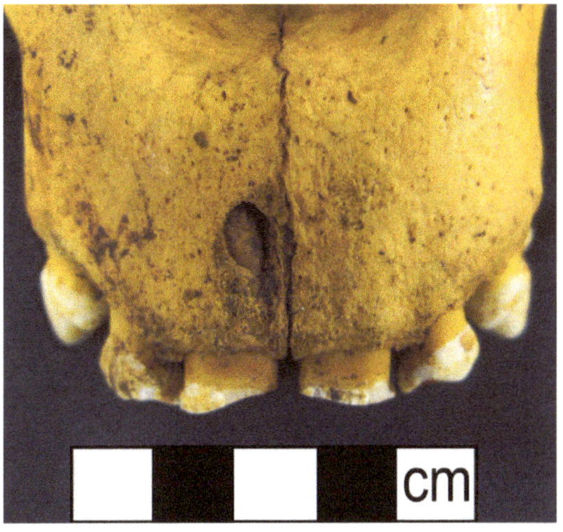

Fig. 1.37 Fistula for a periapical abscess associated with Gilka's right central maxillary incisor

Yolanda (Female: 24 Years)
Yolanda emigrated to the Kasekela community from the Kalande community in 1997, when she was estimated to have been 11 years old. She gave birth to her first recorded offspring (Yamaha) in 1998. Yolanda died in 2006 at the estimated age of 24 years, 3 years after contracting SIVcpz (GSRC). Histopathological analysis of tissue samples taken at necropsy was consistent with end-stage AIDS in humans (Keele et al. 2009). Her skeleton does not exhibit any unusual pathologies for an adult female chimpanzee.

Her skeleton is very complete: all hand and foot bones are present. The hyoid is not preserved, and the sternum is fragmentary (damaged during necropsy). Interestingly, Yolanda has 12 thoracic vertebrae (the modal number for chimpanzees is 13), with 24 paired ribs, and five lumbar vertebrae (mode = 4).

Most of her epiphyses are fused. The acromion and medial clavicle remain entirely unfused. The iliac crest is over half fused, but the secondary ossification center is still separate in some places. The medial margin of the scapula, distal radius, femoral head, and proximal fibula still show a clear epiphyseal line even though the epiphysis is fully fused. The epiphyseal line is completely obliterated in all other cases. The complete adult dentition is in full occlusion and shows no signs of pathology.

Three ribs exhibit an abnormal contour that may be indicative of well-healed fractures. No bony callus remains, so further assessment should rely on radiographs. These ribs (both 12th ribs and another from the right side) also show a moderate level of degeneration of the vertebral articulation.

The proximal ulnar articulation also displays mild to moderate degenerative joint disease in the form of osteophytic growth and proliferative bone on the articular surface. Five of the carpals and tarsals display abnormal, smooth foramina on an articular surface consistent with a subchondral bone cyst. The third and fourth lumbar vertebrae also feature irregular foramina on the anterior vertebral bodies, likely to accommodate larger than average blood vessels.

The shafts of both femora and the left ulna exhibit periosteal reactive bone. Both humeri have osteophytes beginning to bridge the bicipital groove, indicative of partial ossification of the transverse humeral ligament. An intermediate pedal phalanx exhibits a callus of smooth-looking bone near the distal articulation. Its morphology is consistent with benign tumors more commonly found on the cranium. Lastly, two pedal proximal phalanges have unusual nutrient foramen morphology.

Unknown Male (Cranium Only) (20–30 Years?)
The cranium of an unknown male was recovered in the Mitumba community's territory (GSRC). The recovery date is not recorded. This cranium shows no signs of trauma or pathology. The adult dentition was fully erupted at the time of death, though this is estimated from the alveolar morphology as only the second molars remain. There is no evidence for antemortem tooth loss.

Kidevu (Identity Tentative) (Female: 25.6 Years)
The recovery of this skeleton coincides with the disappearance date for Kidevu (GSRC), though the body was not positively identified. Another female chimpanzee

who died within a year of Kidevu's disappearance from the Kasekela community is Gigi. Gigi, however, is estimated to have been nearly 40 years old at the time she died, and the amount of dental wear evident on this skeleton is more consistent with a younger animal (Fig. 1.38). Other females of similar age at death as Gigi include Melissa and Patti. Melissa is estimated to have been 36 when she died and her teeth are much more heavily worn (Fig. 1.89) than this animal's. Patti's teeth are in excellent condition for her age (44 when she died), though she does exhibit antemortem tooth loss, a chipped tooth (Fig. 1.104), and evidence for periodontal disease. I therefore hypothesize that this unidentified animal was Kidevu, who was approximately 25 years old when last seen at Gombe in January of 1992 (GSRC).

The skeleton exhibits some surface abrasion and bleaching from post-depositional damage but is otherwise in good condition and nearly complete, with only a few elements absent. There are 29 carpals/tarsals, nine metacarpals, ten metatarsals, 25 ribs, 65 total phalanges, and one coccygeal vertebra associated with this skeleton. Most of the epiphyses are fused. The medial clavicle is still unfused, and the lateral clavicle is unobservable due to post-depositional damage. Her developmental stage is consistent with it being fully fused. The medial scapular margin and the iliac crest are both more than half fused, and, though fully fused, the sacrum still shows clear fusion sites between the individual vertebrae. The complete adult dentition is in full occlusion. The unfused medial clavicle and incompletely fused scapulae and ilia provide additional evidence that this skeleton is not 40-year-old Gigi.

The left acetabulum has a diffuse, vascularized lesion unassociated with the articular surface. Bone formation pathologies include a discrete callus approximately 3 mm in diameter near the distal articulation of an intermediate manual phalanx, and uneven palmar muscles attachment sites on two proximal manual phalanges, a possible indication of a muscle injury. The posterior aspect of the

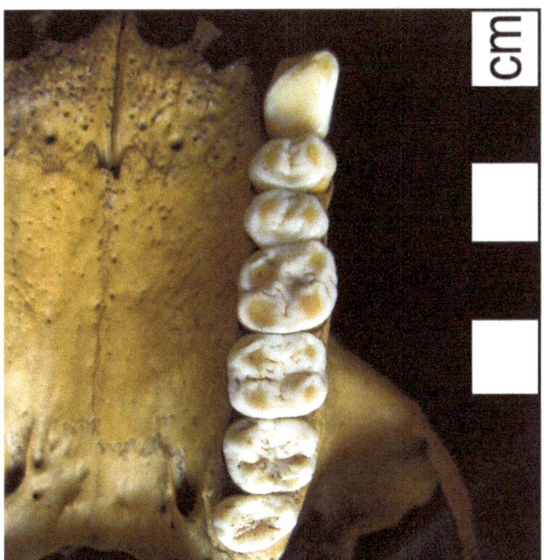

Fig. 1.38 Moderate wear on Kidevu's maxillary dentition is consistent with an animal younger than 40 at the time of death. Note the fourth molar in full occlusion

manubrium shows an area of proliferative bone; the original extent of this lesion cannot be determined due to post-depositional damage.

The left 13th rib exhibits a bifurcated vertebral articulation, indicating extensive remodeling of the costal head and tubercle (Fig. 1.39), though as the corresponding vertebra is not involved, any diagnosis remains tentative. A fourth molar is fully erupted in the left maxilla (Fig. 1.38). A developing alveolus occupies the corresponding place on the left mandible.

Charlie (Male: 25.9 Years)

Charlie was first seen by researchers in June 1963, when he is estimated to have been 12 years old. He and his presumed maternal brother, Hugh, were known for their tandem displays, which helped Charlie in particular rise to dominance. Charlie and Hugh were two of the males most closely involved in establishing the Kahama community, which broke away from Kasekela in 1970. After Hugh disappeared, Charlie was the alpha male of the Kahama community until his death in May 1977. His body was found lying in Kahama stream, a few days after fishermen heard the sounds of fierce chimpanzee fighting in that area. His body had wounds on the head, neck, rump, scrotum, legs, arms, hands, and feet: a pattern of injuries consistent with chimpanzee attack. Charlie is therefore reasonably presumed to have been killed by the Kasekela males during the series of lethal attacks that resulted in the extinction of the Kahama community (Goodall 1986).

Charlie's skeleton is nearly complete, lacking only a few small bones, and the epiphyses are entirely fused, except for the medial clavicle, iliac crest, and sacrum, which are all more than half fused. A clear epiphyseal line is still evident on the acromion, femoral head, and greater trochanter. Charlie's sternum is represented by three body fragments but no manubrium. The hyoid is represented by the body but not the cornua. He lacks any coccygeal vertebrae and one patella. There are 23 carpals/tarsals and 29 phalanges associated with his skeleton. The adult dentition is in full occlusion, and shows little wear (Kilgore 1989).

Unhealed puncture wounds to the right side of his frontal bone (Jurmain 1989) and right maxilla (Fig. 1.40) are further evidence of the manner of Charlie's death. There are four small lesions consistent with subchondral bone cysts evident on wrist and ankle bones in addition to the large lytic lesion on the distal end of his right tibia that is likely related to some kind of bacterial or fungal infection (Jurmain 1989). The paucity of skeletal lesions reinforces the image of Charlie as a healthy, prime-aged male known for his "fearlessness and superb physique" (Goodall 1986, p. 60).

Fig. 1.39 Kidevu's left 13th rib. Note bifurcated vertebral articulation

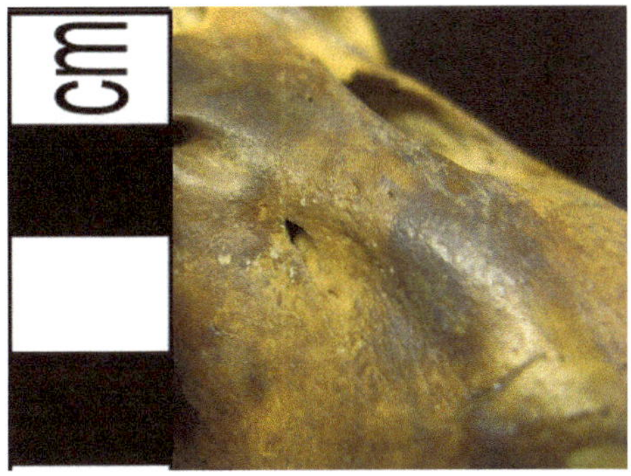

Fig. 1.40 Unhealed puncture wound on Charlie's right maxilla

Pallas (Female: 27.2 Years)

Pallas was first observed in the Kasekela community in March 1965, when she is estimated to have been 9.7 years old (GSRC). As an adult, Pallas was a lower-ranking female whose range was in the middle of the Kasekela community's territory (Goodall 1986). Her first recorded pregnancy, Plato, died in 1970 of a respiratory illness; he is also part of the Gombe chimpanzee skeletal collection (see above). Two other female infants (Villa and Banda) died before reaching 1 year of age in 1975 and 1976. Pallas's final pregnancy, Kristal, died in May 1983 at just under 6 years of age. Kristal's death followed 8 months after her mother's (in September 1982), and is attributed to her being "orphaned" (GSRC). Pallas therefore leaves behind no surviving offspring (Goodall 1986). Pallas's cause of death is the "wasting disease" or gastrointestinal affliction so common among the Gombe chimpanzees (Williams et al. 2008).

Her skeleton is nearly complete and well preserved. There are no phalanges associated with Pallas's skeleton, nor is the body of the hyoid. There are 26 carpals/tarsals preserved, and, interestingly, she has 14 thoracic vertebrae and 28 corresponding ribs. She also has four lumbar vertebrae, suggesting that the 14th thoracic vertebra is truly an additional vertebral element rather than a lumbar vertebra with accessory ribs. The epiphyses are fully fused, with the exception of the acromion, whose end would have included a great deal of cartilage at the time of death (Fig. 1.41). As a similar morphology is seen on Flo (estimated to have been over 50 years old at death), it may be that this is a result of age or shoulder use rather than delayed development. Unfused or partially fused acromion processes have been recorded in humans who vigorously use their shoulders, such as archers and oarsmen (e.g., Stirland 2001). Partial ossification of both transverse humeral ligaments supports the hypothesis that Pallas routinely engaged in vigorous use of her upper limbs.

All other epiphyses are completely fused other than the sacrum, which is more than half fused. The remains of the epiphyseal lines are still evident on the proximal humerus and proximal radius. The complete adult dentition is in full occlusion, and exhibits moderate wear (Kilgore 1989). Pallas exhibits exostoses on both patellae attributable possible muscle injury (Jurmain 1989). She also exhibits osteophytosis on the left transverse process of thoracic vertebra 13 and her left temporal line,

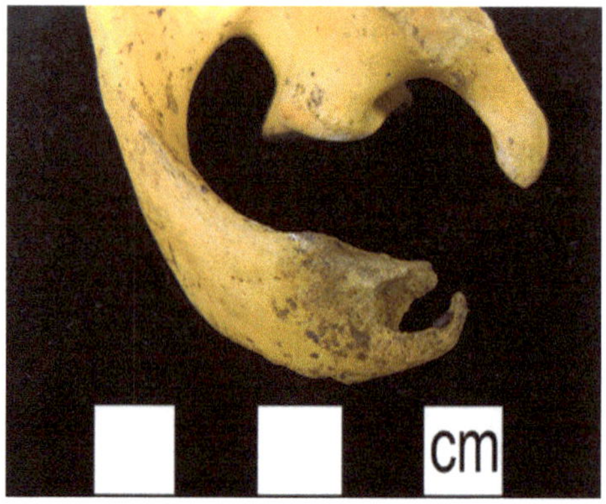

Fig. 1.41 Pallas's unfused acromion, possibly related to forelimb use rather than developmental delay

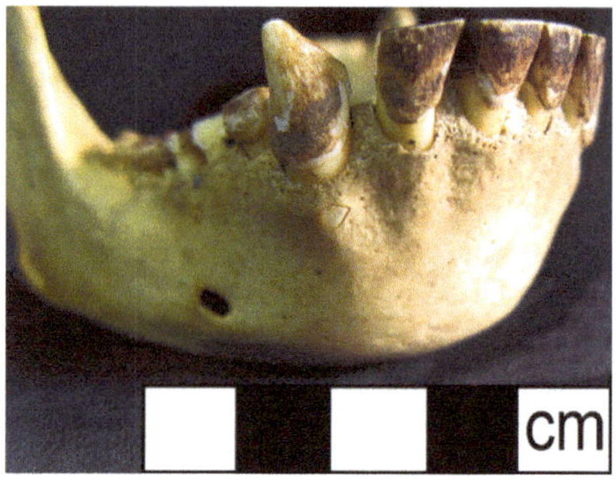

Fig. 1.42 Enamel pearl associated with Pallas's right lower canine

which may also be evidence for muscle injuries/overuse. Periosteal reactive bone is evident on her right first metacarpal. There is a nodule measuring approximately 3 × 3 mm on the labial aspect of the mandibular body, near the location of the root of the right lower canine (Fig. 1.42). It appears to be composed of dental tissue, and the lack of any reactive bone near its location or other evidence for trauma suggests it is not a foreign body. This abnormality may be an enamel pearl, which, though more commonly found on the posterior dentition in humans, can also occur on the roots of incisors and canines (Chrcanovic et al. 2010). There is no skeletal evidence for the severe attack by multiple adult males that Pallas suffered in 1973 in which she sustained a severe gash to her mouth (Goodall 1986).

Rix (Male: 27.38 Years)

Rix was first observed in the Kasekela community in December 1964, when he is estimated to have been 23 years old (GSRC). In November 1968, Rix fell approximately 30 m from a tree and was killed instantly (Williams et al. 2008). Rix is currently an articulated skeleton housed at the University of Dar es Salaam. As a valuable part of the University's comparative collection, the skeleton was observed in its articulated state. The curation of his skeleton makes certain pathological assessments impossible. Rix is therefore excluded from quantitative analysis of skeletal lesions, but this qualitative report provides information on his case.

The third thoracic vertebra is largely reconstructed for the purposes of articulation, but whether damage to this vertebra is due to perimortem trauma or postmortem damage is difficult to determine. His skeleton shows no other signs of perimortem trauma.

Moderately severe inter-proximal caries are evident on the maxillary incisors, between the upper left second premolar and second molar, and between the lower left first and second molars. Rix has just three mandibular incisors. Their even spacing and other morphological features suggests that the right lateral mandibular incisor may have been congenitally absent.

A smooth-edged hole in the left infraspinous fossa measuring approximately 15 mm in length was observed. No reactive bone around the hole was observed, but the thickness of the scapular body as well as the lack of cracking or other damage around the hole makes post-mortem damage unlikely in this case.

Madam Bee (Female: 28.2 Years)

Madam Bee was first observed in the Kasekela community in August 1963, when she is estimated to have been 16 years old. Her daughter, Little Bee, was probably around 3 years old at this time (GSRC). Madam Bee frequently ranged in the southern part of the Kasekela community's range, and transferred to the breakaway Kahama community (to the south) in December 1972. Madam Bee gave birth to her second infant, Honey Bee, in 1965. While Madam Bee had two subsequent pregnancies (in 1971 and 1973), neither of these infants survived past 1 year of age. Because Madam Bee lost the use of her left forelimb during the 1966 polio epidemic at Gombe, it was difficult for her to hold/support an infant, and this may have contributed to their deaths (Goodall 1986).

As part of the efforts of Kasekela males to exterminate the members of the Kahama community, Madam Bee was the victim of a series of attacks towards the end of her life. She sustained a deep gash to one leg after an attack in September 1974, and survived another attack without serious wounds in February 1975. She received a bad wound to her thigh in May 1975, and also fell from a tree while grappling with her assailants. There was at least one unobserved attack in September 1975; Madam Bee was seen with several wounds, the worst of which were on her head and shoulder. The fatal attack occurred a few days after this sighting. Figan, Satan, Jomeo, and Sherry all attacked Madam Bee, dragging, hitting, kicking, stamping on, and pounding her until she could barely move. She eventually dragged herself away. Honey Bee, 10 years old at the time, remained near her mother, "grooming her and keeping off the flies" (Goodall 1986, p. 72), until Madam Bee died, five days after the attack in September 1975. Close to her time of death, Madam Bee was observed with severe wounds on her left ankle, right knee, left wrist, right hand, several wounds on her back, and the hallux of her right foot was almost completely detached (Goodall 1986).

Madam Bee's skeleton lacks both tibiae and fibulae as well as several other, smaller bones. There are eight metacarpals, eight metatarsals, 26 total carpals/tarsals, six cervical vertebrae, 12 thoracic vertebrae, three lumbar vertebrae, 25 ribs, and 51 total phalanges associated with her skeleton. The acromion, medial margin of the scapula, and iliac crest are not yet fully fused. The sacrum still shows obvious epiphyseal lines. The rest of the observable epiphyses are fully fused with obliterated epiphyseal lines. The complete adult dentition is in full occlusion, and shows moderate to severe wear (Kilgore 1989). Her left maxillary lateral incisor was broken antemortem, though no associated abscess is evident. Madam Bee also exhibits alveolar resorption consistent with periodontal disease on the buccal aspect of the mandible at the alveoli for the right first molar and second premolar.

Madam Bee's skeleton exhibits asymmetry attributable to her infection with polio, and the subsequent loss of the use of her left forelimb. The bones of her left forelimb are shorter and lighter than their right-side counterparts (Morbeck 1987). Madam Bee never entirely regained the use of this limb, though limited mobility did return to it (Goodall 1986; Ibid.). The right trochlea also exhibits an area of eburnated articular bone, suggesting the development of degenerative joint disease in her elbow subsequent to preferential use (overuse) of her right limb.

The right fourth metatarsal shows an abscess drainage site, possibly a sequel to trauma. The right fifth metatarsal exhibits a lytic lesion that may have a similar etiology, though the morphology differs in that this element exhibits evidence for healed reactive bone, while the other has smooth margins. Four (4) of Madam Bee's pedal phalanges show asymmetrical muscle attachments on their plantar surface, and one exhibits reactive periosteal reactive bone. One of her manual phalanges shows evidence for a partly-healed fracture.

One of the most dramatic of Madam Bee's traumatic lesions is the non-union fracture to her right ulna, which was only partially healed at the time of death (Jurmain 1989). It is possible she sustained this injury in one of the attacks leading up to her death; the fall from the tree seems a likely candidate for the cause of this injury. A partly healed traumatic lesion to her left ischial tuberosity is also evident (Fig. 1.43), and is likely a result of an attack by chimpanzees (Jurmain 1989). The morphology of the lesion suggests it may have been a puncture from a bite wound.

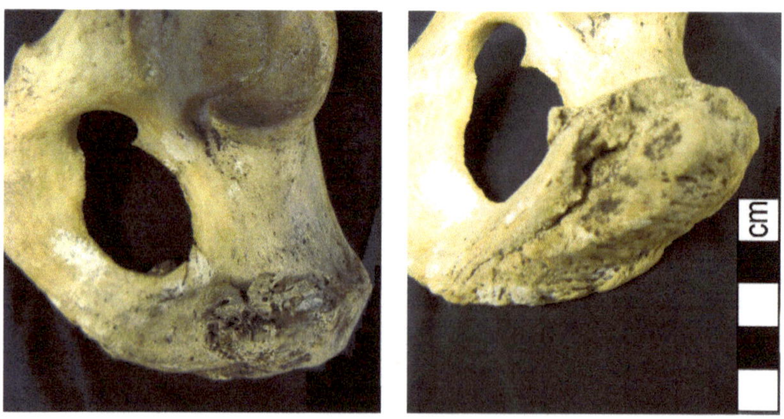

Fig. 1.43 Partially healed puncture wound to Madam Bee's ischium: lateral view (left) and inferior view (right)

There are also several examples of possible perimortem traumata that are consistent with the injuries observed on her body just before her death. This includes an unhealed fracture to the left scaphoid (Fig. 1.44), damage to the medial margin of the left scapula, fraying/crenulation damage to two proximal manual phalanges (Fig. 1.45) as well as similar damage to the distal end of the left first metatarsal (Fig. 1.46). This damage is consistent with wounds inflicted by biting or chewing, which has been observed in cases of infanticide (Arcadi and Wrangham 1999; Kirchhoff et al. 2018) as well as prey consumption (Pobiner et al. 2007). Unhealed fractures to both proximal shafts of the ulnae and a severe fracture to the left proximal radius (Fig. 1.47) may also be evidence for perimortem trauma.

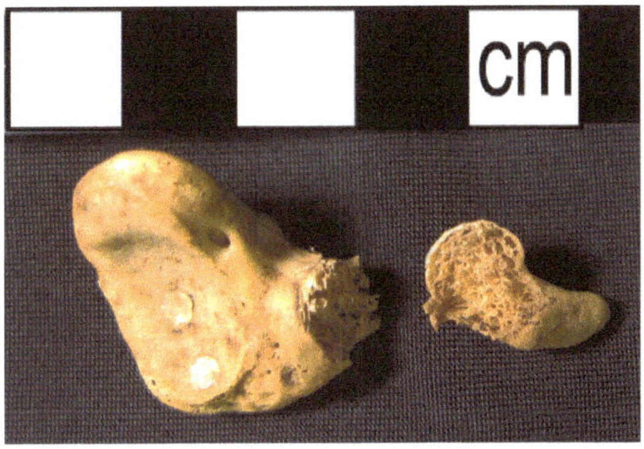

Fig. 1.44 Unhealed, non-union fracture to Madam Bee's left scaphoid

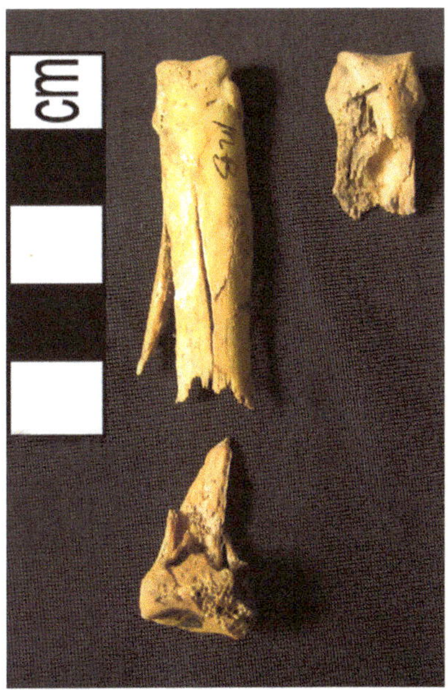

Fig. 1.45 Unhealed crenulated breaks and fraying on two of Madam Bee's phalanges

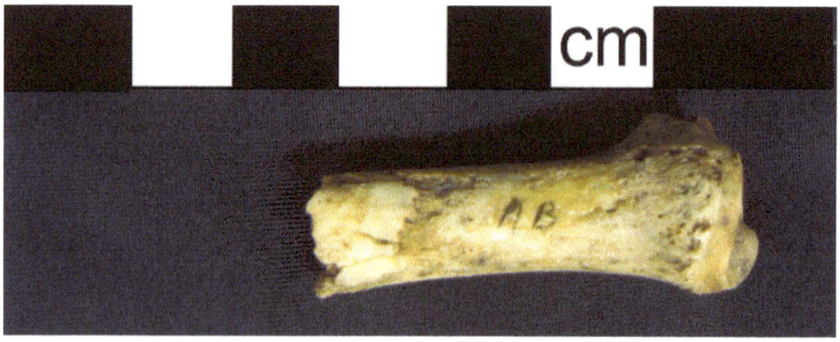

Fig. 1.46 Perimortem damage to Madam Bee's left first metatarsal, consistent with having her great toe bitten off during a fatal attack

Fig. 1.47 Unhealed fracture to Madam Bee's left proximal radius

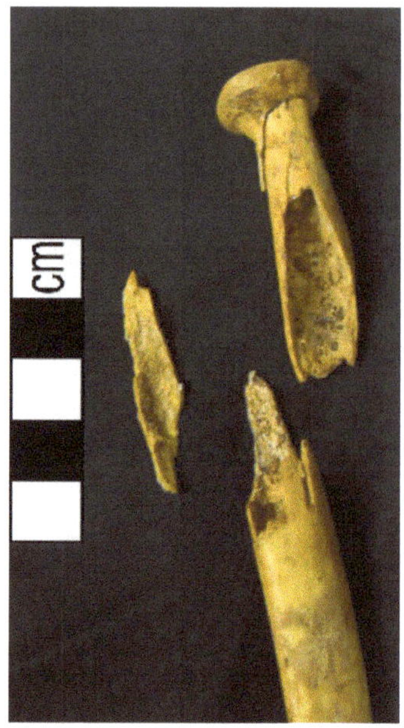

Vincent (Male: 28.5 Years)

Vincent was the alpha male of the Mitumba community until he was badly injured in a fall from a tree. He survived multiple serious traumata, and spent a larger proportion of his time alone than prior to his injury. He re-joined the community after several months, and was killed in December 2004 by the other two adult males of the Mitumba community, Rudi and Edgar (M.L. Wilson unpublished data).

Vincent's skeleton is nearly complete. At least 14 ribs are preserved, but all are fragmentary, having been damaged during necropsy. His clavicles are also fragmentary. His hyoid was not preserved, and 32 carpals/tarsals as well as 44 phalanges are associated with this skeleton. The epiphyseal lines on the medial clavicle, iliac crest, and sacrum are still visible, though the elements are completely fused. The rest of the secondary ossification centers are fused and have obliterated lines. The complete adult dentition was in full occlusion during Vincent's prime, but substantial antemortem tooth loss occurred, likely related to his fall (see below).

Vincent's skeleton exhibits many examples of antemortem trauma. Six ribs exhibit irregular contours and slight calluses indicating relatively well-healed fractures that likely predate his dramatic fall. Three additional ribs exhibit less well-healed fractures (see Fig. 1.48 for one example).

Both lateral and the left central maxillary incisors are chipped on the incisal surface. The pulp cavity of the right maxillary canine is exposed. The chipped surfaces of the incisors are smooth: the damaged surfaces exhibit wear. The left canine is much less worn, suggesting that exposure of the pulp cavity may have been traumatic in nature rather than through normal wear, especially as the left maxillary canine is in good condition. Three abscess drainage sites associated with the right canine are evident on the maxilla (Fig. 1.49). The right zygomatic arch displays a partially healed non-union fracture (Fig. 1.50). The mandible exhibits the antemortem loss of

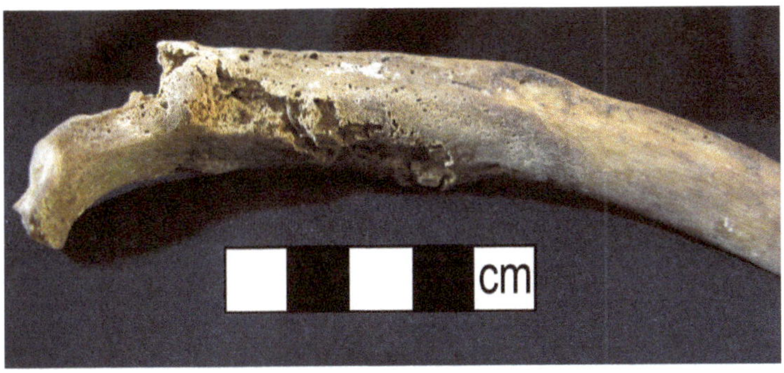

Fig. 1.48 Partially healed fracture to vertebral end of one of Vincent's ribs

Fig. 1.49 Multiple fistulae associate with periapical abscesses in Vincent's maxillary dentition

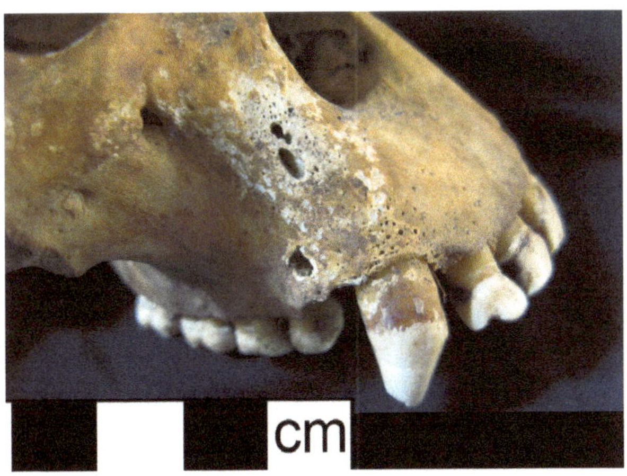

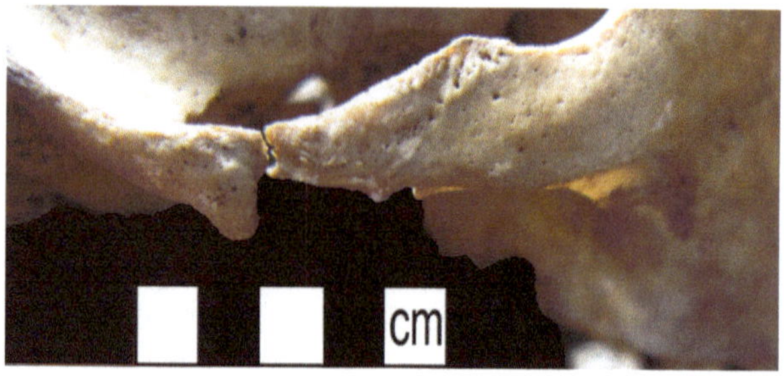

Fig. 1.50 Non-union fracture to Vincent's right zygomatic arch

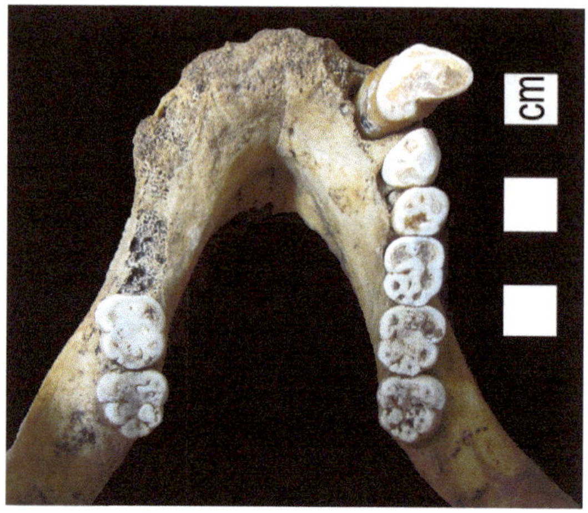

Fig. 1.51 Antemortem tooth loss in Vincent's mandible

the left first molar through the right lateral incisor (Fig. 1.51). The alveoli are mostly closed, though a fragment of the broken left canine remains in situ. The left mental foramen is enlarged, probably to serve as a drainage site related to periodontal infection. The left canine fragment and the moderate wear on the surviving dentition support the hypothesis that these teeth were lost due to injury rather than periodontal disease. (Though some carious lesions are recorded, see below.)

The fifth sacral vertebra is deviated compared to the rest of the sacrum, and reactive bone indicates a healing fracture (Fig. 1.52). The left ischium exhibits a non-union fracture separating the ischial tuberosity from the rest of the os coxa (See Fig. 1 in Terio et al. 2011). Other partly healed fractures include the body of the first lumbar vertebra, the distal left fibula and tibia, and the left and right calcanei (Fig. 1.53). New, waxy-looking bone laid down near the right internal acoustic meatus (Fig. 1.54) may be related to cranial trauma from the fall.

Perimortem, unhealed traumata also abound on this skeleton. Many of the wrist and ankle bones are incomplete or punctured. In particular, the left calcaneus and talus are incomplete, the left distal fibula is missing, and the left distal tibia is punctured. Several metapodials have exposed trabecular bone on the proximal articulations. Unlike in degenerative joint disease, the margins of these lesions are sharp, indicating a possible traumatic origin. The left scapular blade, right iliac crest, and

Fig. 1.52 Deviation of Vincent's fifth sacral vertebra indicates a likely fracture

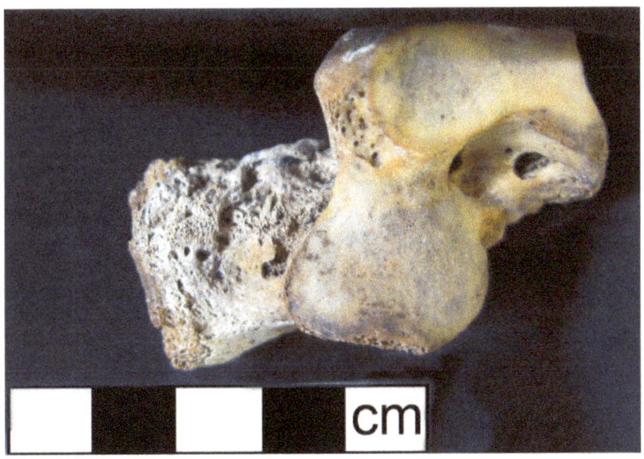

Fig. 1.53 Partially healed compression fracture on Vincent's calcaneus

the spinous process of the fourth lumbar vertebra show un-healed fractures. Two of the ribs may show perimortem step fractures, as inferred from the direction of the break (towards the concavity of the rib), though this is difficult to separate from postmortem damage to the rib cage during necropsy to access the thoracic cavity (though these fractures would most likely be towards the convexity of the rib). The left maxilla shows an un-healed puncture wound (Fig. 1.55).

Damage to Vincent's skeleton is consistent with recorded behaviors during chimpanzee agonistic encounters. When chimpanzees attack, the face, hands, and feet of their targets are often bitten, and the attackers may jump on or strike the back of the

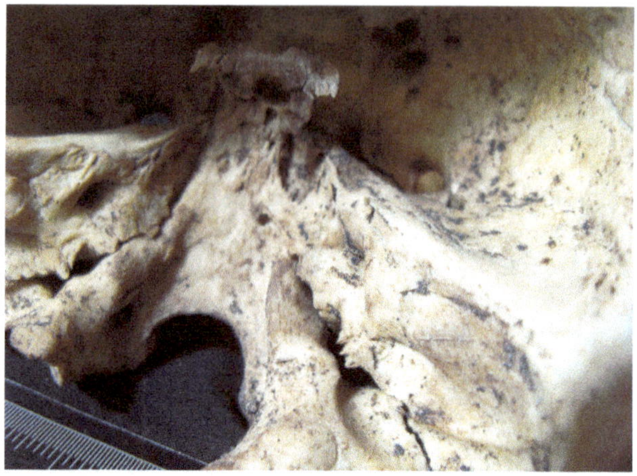

Fig. 1.54 Abnormal bony growth near Vincent's right internal acoustic meatus

Fig. 1.55 Perimortem puncture wound to Vincent's left maxilla

target (targets often assume a crouching position in an attempt to protect themselves) (Goodall 1986; Wilson et al. 2004). Damage to Vincent's scapula, ilium, vertebra, and possibly ribs is consistent with blows to the dorsum of the body. Damage to Vincent's left ankle is consistent with bite wounds to this region and are comparable to skeletal damage to the first metatarsal of Madam Bee, an adult female chimpanzee from Gombe whose great toe was bitten off during a fatal attack (see above) (Goodall 1986). Extensive soft tissue damage to Vincent's face, including removal of his nose, was noted at necropsy (the right testis was also absent) (GSRC data). This is consistent with the canine puncture evident in Vincent's left maxilla.

Interproximal caries, most of which are only moderately invasive, are evident on the right maxillary first premolar through the left lateral incisor. The first premolar has more extensive involvement than the other teeth. Relative to the right temporal line, the left temporal line is deviated inferiorly on its posterior aspect. Both coronoid processes (the insertion point for the temporalis muscle) show osteophytic growths, which may indicate partial ossification of the muscle's tendon. This may be related to the mechanical alterations to mastication that likely accompanied the sudden loss of several teeth. The right femur shows extensive remodeling of the lesser trochanter, possibly related to relying more heavily on this limb while recovering from the severe fracture to the left ischium, which, as the origin for the hamstrings, would have made use of this muscle group very difficult.

Other bone formation pathologies include periosteal reaction on the right tibia and an intermediate manual phalanx. Two other manual phalanges show uneven palmar muscle attachments, and a third intermediate manual phalanx exhibits a protuberance of unknown etiology (~3 × 4 mm). Both bicipital grooves have bridging osteophytes indicating partial ossification of the transverse humeral ligament. Lastly, one metacarpal exhibits both bone formation and bone loss related to the same pathology: a fistula approximately 3 mm in diameter accompanied by surrounding remodeled bone (Fig. 1.56).

Of special interest in this case is the degree of healing evident on the antemortem traumata because the time between these injuries and death is known (3 months). In most cases, new bone still has a woven appearance and callus formation is incomplete.

Injuries including falls (8% of male deaths) and conspecific killings (up to 39% of male deaths) are both relatively common causes of death at Gombe (Williams et al. 2008). The most common cause of death is illness (Ibid.), but falls and conspecific

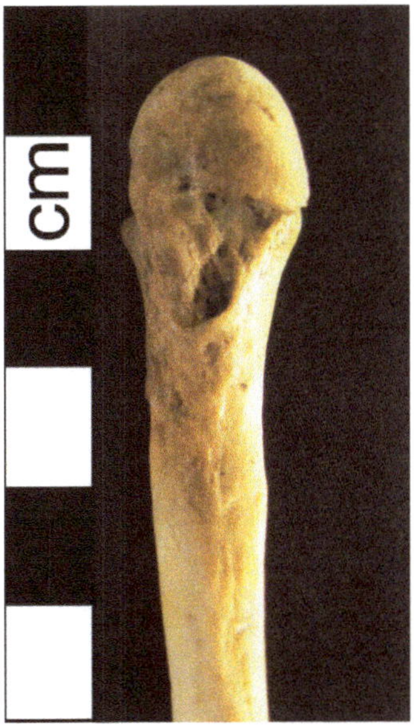

Fig. 1.56 Vincent's metacarpal exhibiting both pathological bone loss and bone formation

aggression are more likely to leave evidence on the skeleton (Kirchhoff 2010, see also Chap. 2, this volume).

Winkle (Cranium Only: Identity Tentative) (Female: 28.9 Years)
Recovered from the Kasekela community's range in 1988, this cranium's discovery coincided with the disappearance of adult female Winkle. Winkle was first observed in the Kasekela community in August 1968, when she is estimated to have been 9 years old. Her first recorded pregnancy was Wilkie, a son born in October 1972, who remained in the Kasekela community until his disappearance in January 2013. Her subsequent offspring were born in December 1978 (Wunda, female) and December 1984 (Wolfi, male). Both disappeared in May 1990 (GSRC).

Only Winkle's neurocranium remains; most of the facial bones were broken postmortem, though portions of both orbits survive. The left orbit shows evidence of postmortem rodent gnawing on its medial aspect (Fig. 1.57).

Both parietals and the occipital show multiple bone loss pathologies in the form of small, regular divots with smooth margins. The etiology of this pathology is unknown, though similar lesions observed in the fossil record have been reported as consistent with well-healed depression fractures (Novak and Hatch 2009; Indriati and Antón 2010). These lesions may also be consistent with the healed signs of infection (Ortner 2003). Radiographic analysis may be helpful for future diagnosis (Galloway and Wedel 2014).

Satan (Male: 29.8 Years)
A large male from the Kasekela community, Satan is represented by only a few skeletal elements: the mandible, both ossa coxae, both humeri, and the left ulna, whose distal end is missing due to postmortem damage. He died in 1987 of a respiratory infection (Williams et al. 2008). Satan exhibits osteophytic growth across the

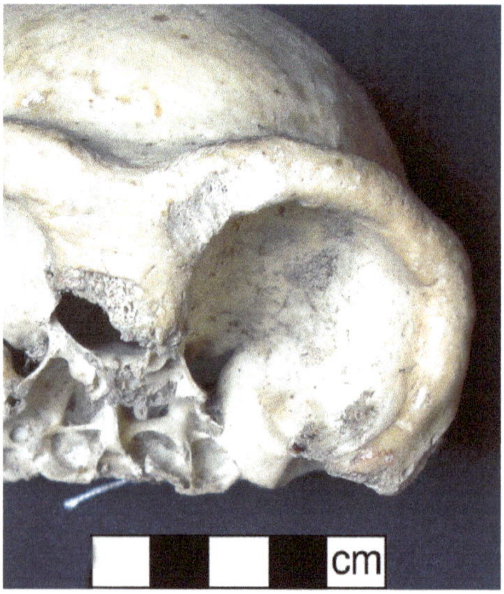

Fig. 1.57 Rodent gnawing on Winkle's left orbit

intertubercular groove of his right humerus, indicating partial ossification of the transverse humeral ligament.

Satan was first observed in January of 1965, when he was estimated to have been 7.5 years old. He was still dependent on his mother, Sprout, at that time. Satan is her first known offspring, but given that she is estimated to have been 24 years old in 1965, she may have had one or more earlier pregnancies unrecorded by human researchers (GSRC).

Tumaini (Male: 30 Years)
Tumaini (alternate spelling: *Tumaine*) is an identified male from the unhabituated Kalande community, whose home range extends through the southern part of Gombe National Park. Tumaini was killed by poachers in 2002, and his body was found partially burned and with multiple machete wounds (GSRC unpublished data). Unhealed skeletal traumata reflect his cause of death. Most of the axial skeleton (ribs and vertebrae), many hand and foot bones, and the left lower extremity are not preserved, and are likely related to the manner of his death (and subsequent partial recovery of possibly scattered remains). His sternum is not preserved, both clavicles and scapulae are fragmentary, and only three rib fragments are associated with his skeleton. Tumaini has two lumbar vertebrae, but no other pre-sacral vertebrae. The hyoid is represented by the body only. The right os coxa is present, but fragmentary, while the left is not preserved. The right femur and tibia are present, but no patellae or fibulae. There are 18 carpals/tarsals, six metacarpals, six metatarsals, and 33 phalanges associated with his skeleton.

Most secondary ossification centers are fully fused. The iliac crest is still unfused, and several epiphyses still have a visible line on the fused epiphysis: the medial scapular margin, proximal humerus, proximal radius, distal radius, distal ulna, femoral head, proximal tibia, and sacrum. The complete adult dentition is in full occlusion, with no antemortem tooth loss.

Perimortem traumata include spiral fractures to both clavicles and damage to the scapular blades and right ilium (Fig. 1.58). The three preserved rib fragments also exhibit unhealed spiral fractures. Several of the anterior teeth are calcined, indicating that he was at least partly burned by the poachers.

Antemortem traumata include a pedal proximal phalanx with a missing distal end. Reactive bone indicates that some remodeling occurred between the time of this injury and death, even though the margins of the break are still relatively sharp. Three metapodials exhibit large, mid-shaft calluses that dramatically alter the normal contour of the bone (Fig. 1.59). A manual proximal phalanx is similarly affected, though its degree of healing is not as extensive (Fig. 1.60).

The first sacral vertebra exhibits mild osteophytic lipping. The fourth lumbar vertebra does not. Bone loss pathologies include notches on the distal aspect of both maxillary canines (Fig. 1.61), likely resulting from wear rather than carious lesions due to their morphology and the notches' disassociation with other tooth surfaces. The left maxilla has an abscess drainage site related to the canine. The left maxilla also has a bone formation lesion (Fig. 1.62). The degree of relief and circle-shape suggests it may be an osteoma, though the irregular contour may be indicative of a

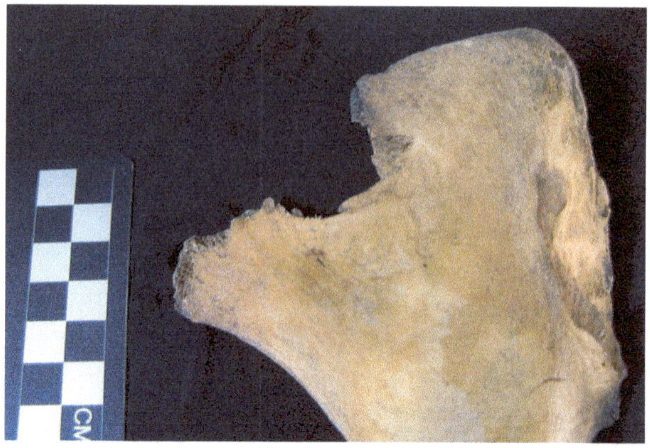

Fig. 1.58 Damage to Tumaini's right ilium

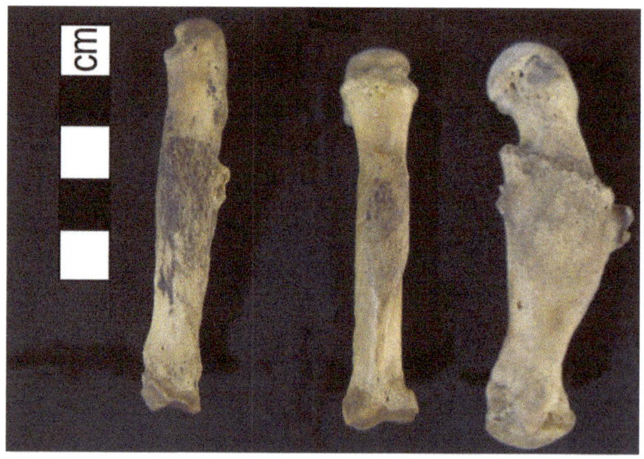

Fig. 1.59 Three of Tumaini's metapodials with healing fractures

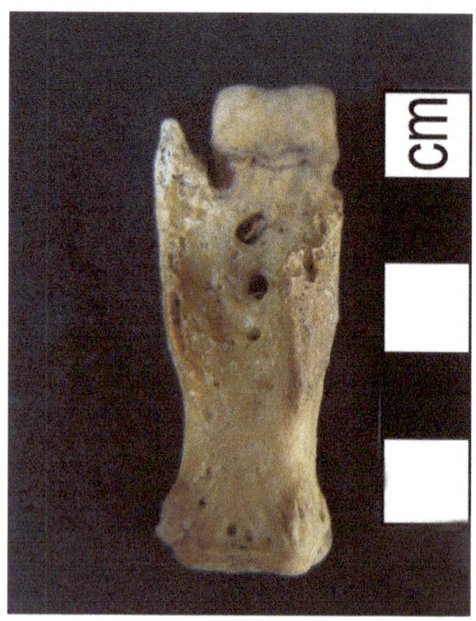

Fig. 1.60 Partially healed fracture to one of Tumaini's manual phalanges

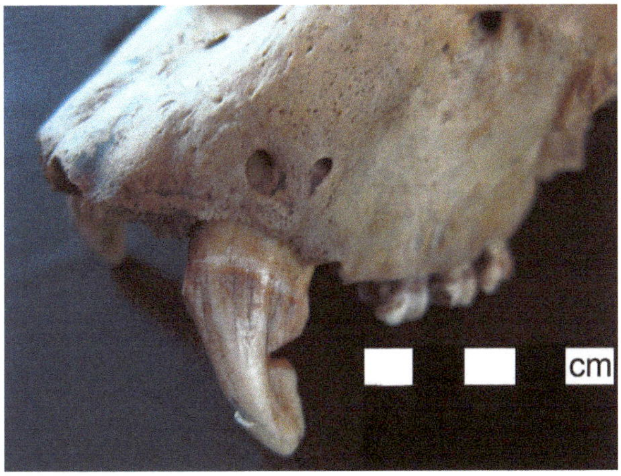

Fig. 1.61 Notch on the distal aspect of Tumaini's upper left canine. Note associated abscess drainage sites

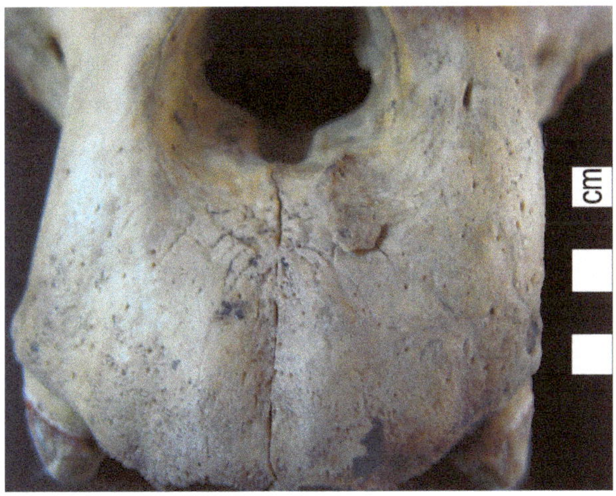

Fig. 1.62 Bone formation pathology on Tumaini's maxilla

traumatic origin for this lesion. The maxilla is also very asymmetrical, with the left side more bulbous than the right. Whether the asymmetry is related to the bone formation lesion or the abscess is unclear; radiographic analysis in future may be informative.

Alveolar erosion is evident near the mandibular left first premolar as well as the maxillary canines, and the maxillary left second premolar is chipped. Lesions are present on both mandibular condyles (Fig. 1.63); the mandibular fossae are not involved. The smooth margin of these lesions is consistent with the morphology of a subchondral cyst.

The left temporal line is positioned more laterally in its posterior aspect than the right (Fig. 1.64). Periosteal reaction is evident on the right tibia and an intermediate manual phalanx. Proliferative lesions on a metatarsal and the right ulna are also likely to have resulted from periosteal response. Two manual proximal phalanges

Fig. 1.63 Possible subchondral cyst on Tumaini's mandibular condyle

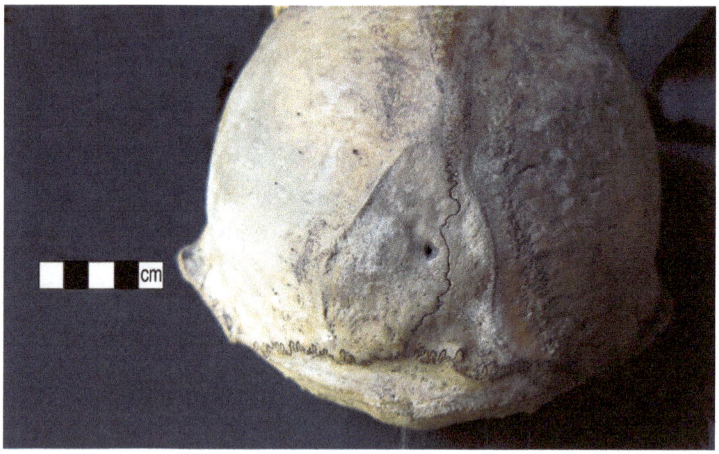

Fig. 1.64 Deviation of Tumaini's left temporal line

have asymmetrical palmar muscle attachments, and one toe has fused intermediate and distal phalanges, possibly as a result of trauma.

Passion (Female: 30.6 Years)

Passion, a high-ranking female from Kasekela, is most famous for killing and eating several young infants in her own community. With her daughter Pom's assistance, Passion may have killed up to ten infants between 1974 and 1977. Three of these infanticides were observed (Goodall 1986). Passion died in February 1982 of a "wasting disease"/gastrointestinal illness (Williams et al. 2008).

Passion was first observed in December 1964, when she is estimated to have been 13 years old. Her first observed pregnancy, also in 1964, was an infant of unknown sex who disappeared within a week of birth. Pom was Passion's second recorded pregnancy. Pom was born in July 1965, and emigrated to the (at the time, unhabituated) Mitumba community in 1983, after her mother's death. Gombe researchers hypothesize that Pom died in 1987, as she was sighted at Mitumba in 1986, but not in 1988. Passion's third recorded pregnancy was her son, Pax, born in 1977. He is suspected to have sustained testicular injury/loss during an episode of intercommunity violence between Kasekela and Kahama, supported by his overall small size and poor scrotal development. He remained a lower-ranking male in the Kasekela community until his death in September 2015 (GSRC).

Passion's skeleton is mostly complete. There are 18 carpals/tarsals attributed to her skeleton, 24 ribs, one patella, and 82 total phalanges. The right metatarsals and right fibula were not recovered. All of the (observable) epiphyses are fused except for the medial margin of the scapula. The lines of the fused epiphyses are still evident on the iliac crest and distal fibula.

The complete adult dentition is in full occlusion, and displays moderate wear. Loss of alveolar bone indicating periodontal disease is evident in the posterior dentition (Kilgore 1989). The maxillary right first molar was chipped on the buccal aspect prior to death as indicated by the wear on the break surface. Loss of enamel and dentin on the maxillary incisors is consistent with interproximal caries (Fig. 1.65).

A traumatic dislocation of both 13th ribs is indicated by the remodeling of both rib heads (Fig. 1.66) and the corresponding vertebra (Jurmain 1989). Uneven muscle attachments and reactive bone near attachment sites on three manual phalanges are indicative of severe muscle pulls or strains. She exhibits healed fractures to an intermediate pedal phalanx, two manual proximal phalanges, and the left fifth metatarsal.

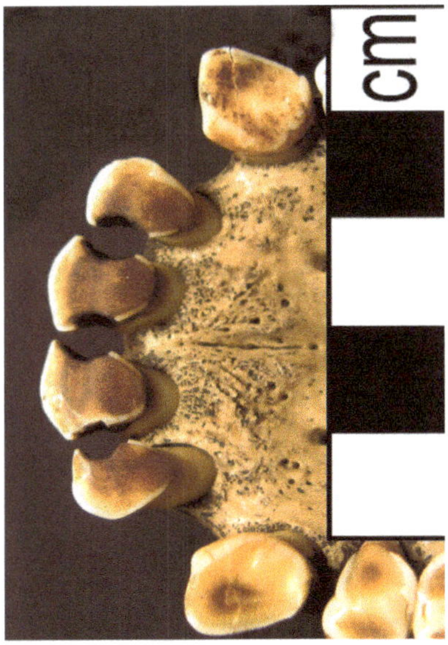

Fig. 1.65 Interproximal caries on Passion's maxillary incisors

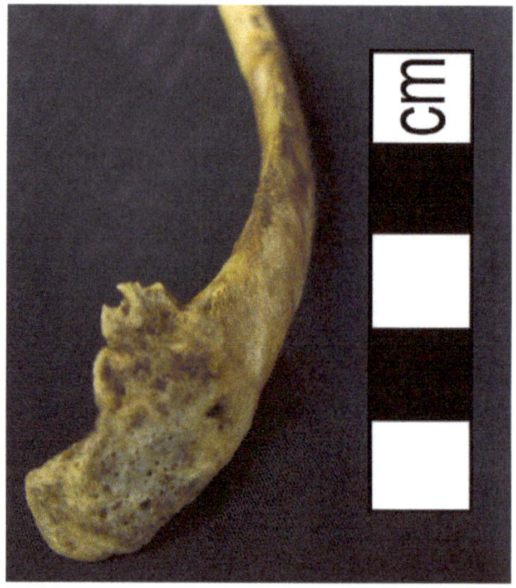

Fig. 1.66 Passion's rib head, remodeled due to a dislocation injury

On two pedal rays, the intermediate and distal phalanges are fused. Passion also has a patent olecranon foramen on her right humerus. Both acromion processes exhibit pitting on their inferior surfaces that, in humans, is consistent with repetitive overhand motions (Neidich 2014; Roberts et al. 2007; Miles 1994).

Passion has patches of periosteal reactive bone on one pedal and one manual phalanges, the distal shaft of the right ulna, the anterior shaft of both femora, and the medial shaft of the left tibia. Associated with her skeleton is an unidentified fragment of reactive bone (approximately 20 mm in length) whose provenance is unknown. It is possible this represents heterotopic ossification subsequent to an injury (Barfield et al. 2017). This is difficult to determine for Passion since the fragment is not closely linked to any particular body region.

A depression measuring approximately 9 mm in diameter near bregma (Fig. 1.67) is consistent with the morphology of depression fractures to the cranial vault recorded in other contexts (Novak and Hatch 2009; Jurmain 1989, 1997). Heterotopic ossification is often associated with head injuries in humans (Barfield et al. 2017). Even in the closely observed Gombe chimpanzees, injuries are often not directly witnessed by observers, and there are no records of an observed head injury for Passion.

Jomeo (Male: 30.9 Years)

A male of impressive size, Jomeo lost interest in acquiring high dominance rank during adolescence, an unusual occurrence for a male chimpanzee. Jomeo was noted to have sustained severe wounds at approximately 9 years of age, and it may be that whatever experience resulted in these wounds precipitated this lack of interest (Goodall 1986). Jomeo was first observed in November 1964, when he is estimated to have been about 8.5 years old (GSRC). His mother, Vodka, and brother, Sherry, were both regularly observed starting in 1967 (Goodall 1986), when Sherry

Fig. 1.67 Possible healed depression fracture on Passion's cranium, near Bregma

is estimated to have been 5.5 years old (GSRC). Jomeo and Sherry associated closely until Sherry's death, and Jomeo regularly supported his brother during dominance displays (Goodall 1986).

Sherry disappeared in 1979, a suspected victim of intercommunity violence (GSRC), at which point Jomeo increased his association with then-alpha Figan, and began to associate more frequently with unrelated adolescent males. The orphan Beethoven (tentatively identified as part of the Gombe chimpanzee skeletal collection) frequently associated with Jomeo, and Jomeo regularly tolerated the presence of other young males following him (Goodall 1986).

Satan is often credited with being the largest male chimpanzee observed at Gombe, and data on body mass collected during their lives indicate that Satan and Jomeo were similar in size (Pusey et al. 2005), but Jomeo's skeleton is markedly larger; the length of Jomeo's right os coxa is 20 mm greater than Satan's. It is interesting that a male of such impressive size showed comparatively little interest in climbing the dominance hierarchy. Both his innate personality as well as early experiences (e.g., the severe injuries he sustained around age 9) may have contributed to Jomeo's behavior. Despite his lack of participation in intracommunity aggression, Jomeo regularly took a prominent role in violent intercommunity encounters, as in the cases where the males of his home community (Kasekela) attacked members of the break-away Kahama community (Goodall 1986).

Jomeo's skeleton is nearly complete, lacking only a few hand/foot bones. There are 23 carpals/tarsals, five metatarsals, and 44 total phalanges associated with his skeleton. The epiphyses are all fully fused, except the transverse processes on the first lumbar vertebra, which are incompletely fused (Fig. 1.68).

The epiphyseal line is still visible on the medial clavicle. The fourth lumbar vertebra has fused to the sacrum. The complete adult dentition is in full occlusion, and exhibits moderate wear.

The upper left canine was damaged sometime during life (Fig. 1.69). Little wear on the broken edges of the tooth indicates the possibility that this occurred not very

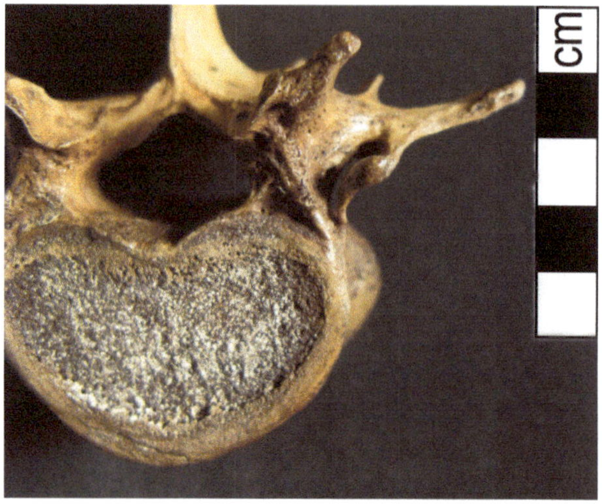

Fig. 1.68 Incompletely fused transverse process on Jomeo's first lumbar vertebra

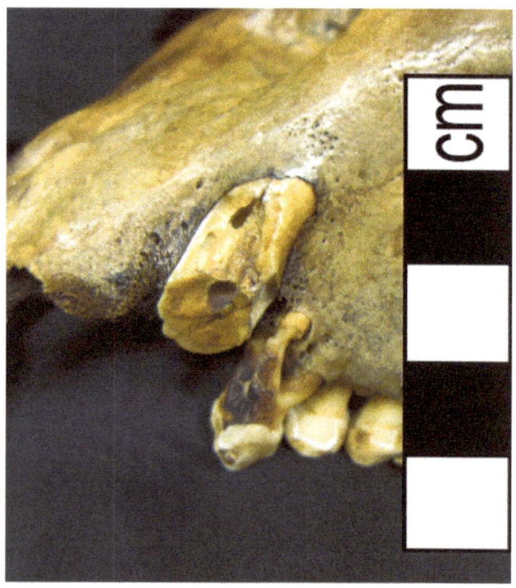

Fig. 1.69 Jomeo's left maxillary canine; the breakage may have a traumatic origin

long before death, but bony evidence as well as observational data make this unlikely. Tooth breakage that exposes the pulp cavity often results in the formation of an abscess near the tooth. If sufficiently purulent (and untreated), the abscess will form a fistula through the alveolar bone so that it can drain. If such a fistula formed for Jomeo's broken canine, it must have remodeled by the time he died (indeed, there is a patch of abnormally porous bone in the periapical region of his maxilla that may correspond to just such a drainage site). In addition, during life, Jomeo was observed to have developed a dental abscess (which burst and drained) on the left side of his face (Goodall 1986). While observers initially attributed the infection to one of Jomeo's upper molars, the skeletal evidence suggests that the maxillary left canine was the injured tooth. Antemortem loss of the mandibular premolars on the left side may be associated with this injury, and provides additional support for the canine breaking antemortem. Though the alveoli for the shed mandibular premolars are well healed, the enlarged left mental foramen, a common pus drainage site in cases of infection of the mandibular dentition, supports the traumatic tooth loss hypothesis. These dental injuries occurred after the eye injury Jomeo sustained in 1978, which resulted in at least partial blindness in that eye, and distinctive, light-colored scar tissue on the globe (Goodall 1986). In addition, Jomeo exhibits loss of dental tissue on his right maxillary lateral incisor, left maxillary central incisor, and left maxillary first premolar that is consistent with dental caries.

A left rib, distal phalanx, and the distal shaft of the right ulna (Fig. 1.70) all show calluses indicative of well-healed fractures. The fracture to the right ulna was an extensive, severe injury, based on the size of the remaining callus (over 3 cm). The fracture is well-healed in that no reactive bone is evident on the ulna. Two additional manual phalanges have bone formation lesions that may represent less severe and/or more thoroughly healed fractures. Other minor lesions include periosteal reaction on two manual phalanges, and uneven muscle attachments on one manual and one pedal proximal phalanx.

A more dramatic lesion is the lesion evident on Jomeo's right greater trochanter (Fig. 1.71). The reactive bone may indicate a localized infection.

Patches of eburnation are evident on both elbows (Fig. 1.72). This may be a sequel to his ulnar fracture. Osteophytes on the anterior surface of the left patella indicate partial ossification of the quadriceps tendon in this region, possibly due to a severe muscle pull or strain.

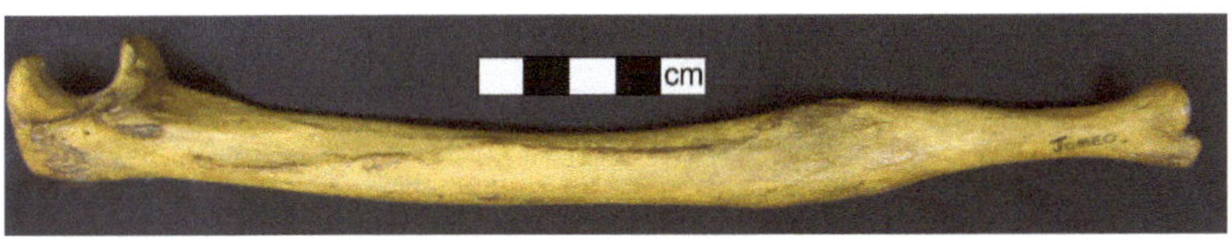

Fig. 1.70 Healed fracture to the distal shaft of Jomeo's right ulna

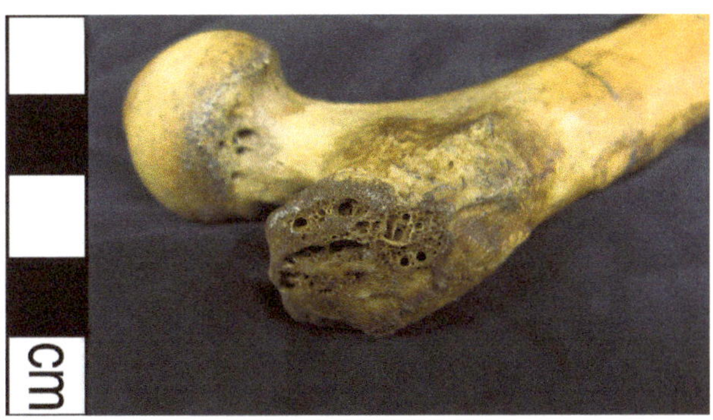

Fig. 1.71 Reactive bone on Jomeo's right greater trochanter

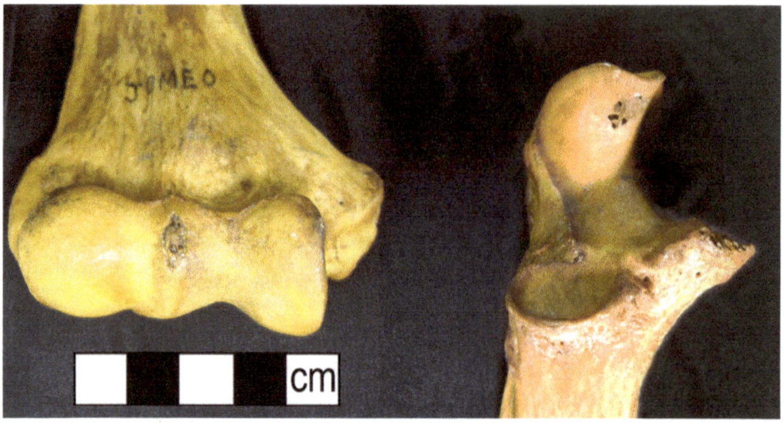

Fig. 1.72 Eburnation on Jomeo's distal right humerus and proximal right ulna

Whether any of these skeletal lesions are related to the severe injuries Jomeo sustained around the age of nine remains speculative. Severe injuries during adolescence may still show signs into adulthood, though being skeletally immature at the time of injury would have resulted in increased bone remodeling subsequent to the injuries compared to injuries sustained as an adult.

Miff (Female: 30.9 Years)

Miff was first observed in August 1963, along with her older brother, Pepe, younger brother, Merlin, and mother, Marina. Marina is estimated to have been 37 years old at the time, so it is likely she had offspring prior to Pepe. Miff is estimated to have been 7 years old in 1963 (GSRC). When Marina died in 1965, Miff took care of Merlin, allowing him to share her nest, and waiting for him while traveling. Merlin, however, never recovered from being orphaned, and died during the 1966 polio epidemic (Goodall 1986).

Because of her early experiences caring for her brother, Miff proved to be a competent mother (Ibid.). Despite this, several of her offspring did not survive to adulthood, including her son Michaelmas (1973–1986, wasting disease), and her son Mel (1984–1994, possible intracommunity attack) (GSRC), both of whom are also part of the Gombe chimpanzee skeletal collection (see above). Miff's daughter, Mo, died in August 1985 of a respiratory infection, at age 7. Miff's firstborn, Moeza, was born in January 1969, emigrated to the Mitumba community in 1983, and died of a respiratory infection in 1996, at age 27 (GSRC). Neither of Moeza's offspring survived to adulthood (GSRC).

Miff herself died in 1987 of a respiratory infection (Williams et al. 2008). Her skeleton is almost entirely complete, lacking only the hyoid, coccyx, and some phalanges (there are 43 total phalanges associated with her skeleton). Most of the epiphyses are fully fused. An epiphyseal line is still evident on the medial clavicle and proximal humerus. The iliac crest and sacrum are more than half fused.

The acromion remains unfused, but see the section on Pallas for comments on this phenomenon. The complete adult dentition is in full occlusion and exhibits mild to moderate wear. Several teeth were lost antemortem: the upper left first premolar, all of the upper left molars, the upper right third molar, and the lower left canine (Fig. 1.73). The right mandibular molars are more worn than the left, likely due to the fact that the upper left molars were missing. Given the overall wear stage of Miff's teeth and the lack of carious lesions on teeth adjacent to the lost teeth, it is possible the tooth loos exhibited in her case is traumatic in origin. Lytic lesions as well as postmortem damage of the alveolar bone make this diagnosis unsure.

The left deltoid tuberosity exhibits severe remodeling and extra bony growth, indicating likely trauma to this muscle (Fig. 1.74). It is not clear whether this injury is a result of the attack on Miff by multiple males in 1977. Her son, Michaelmas, was seized and displayed with (Goodall 1986), resulting in a hip dislocation injury

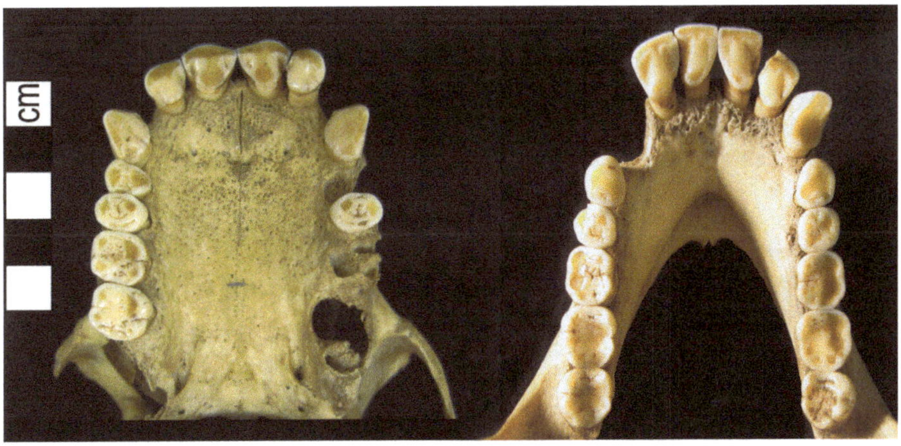

Fig. 1.73 Miff's maxillary (left) and mandibular dentition (right)

Fig. 1.74 Bony evidence for a likely injury to Miff's left deltoid

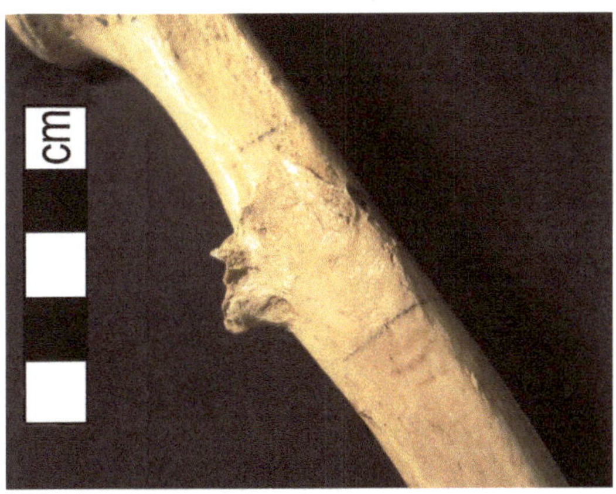

(Jurmain 1989). Miff was often the target of violent attacks during her life, particularly while she was cycling (Goodall 1986), and the injury may have been sustained during some other attack, or during a fall.

Two manual phalanges exhibit calluses consistent with well-healed fractures. Both mandibular fossae (Fig. 1.75) met the criteria for TMJ osteoarthritis (Rando and Waldron 2012), which may, in Miff's case, be related to her antemortem tooth loss.

Atlas (Male: 31.3 Years)

Atlas presents a fascinating case, as he is a chimpanzee who has been studied since his birth in September 1967 until his death in January of 1999. His mother, Athena, was first observed in the Kasekela community in May 1965, when she is estimated to have been 12 years old. Atlas is her first recorded pregnancy, and very likely her first live birth. Athena's second infant arrived in 1973. Aphro (Atlas's sister) emigrated to the Mitumba community in 1988, and is currently (in 2019) the oldest adult female in that group, at 46. One of Aphro's daughters, Andromeda (Atlas's niece), an infant killed by males of the Kasekela community in 2005, is also part of the Gombe skeletal collection (see above) (GSRC).

Atlas's death in 1999 is attributed to a "wasting disease" that is a relatively common cause of death for chimpanzees at Gombe, and is likely related to intestinal parasites in his case (Williams et al. 2008). His skeleton is very complete, lacking only a few hand/foot bones. There are 29 total carpals/tarsals and 52 total phalanges associated with his skeleton. His skeleton is also well preserved (even the hyoid is present), with most secondary ossification centers fully fused. The medial clavicle remains unfused, and the iliac crest and sacrum are more than half fused. Fully fused epiphyses that still show a clear epiphyseal line include the proximal humerus, proximal radius, and proximal ulna. The complete adult dentition was in full occlusion prior to antemortem tooth loss (see below).

Observed traumata include a chipped maxillary right lateral incisor, and three ribs with abnormal contours suggesting well-healed fractures. Three metapodials

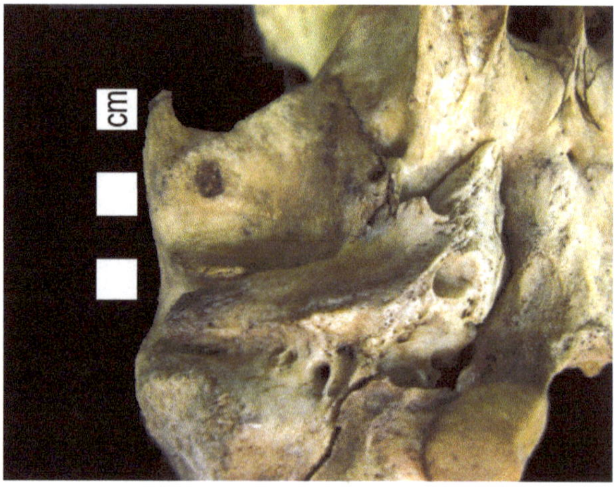

Fig. 1.75 Eburnation on Miff's right mandibular fossa

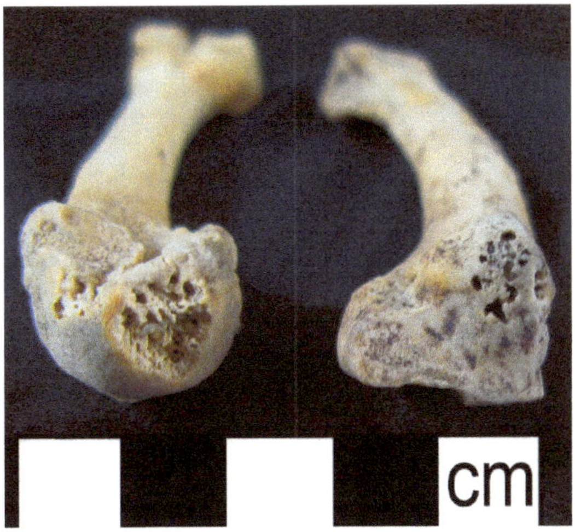

Fig. 1.76 Examples of Atlas's metapodials with severe erosional lesions to the joint surface

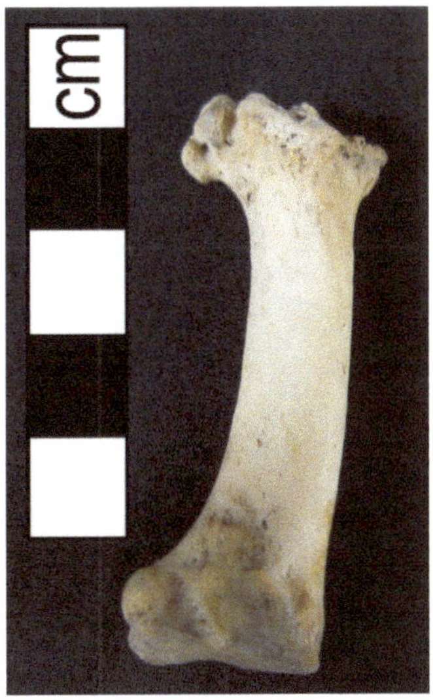

Fig. 1.77 Atlas's left fourth metatarsal, exhibiting severe remodeling of the distal articulation

show severe degeneration of a joint surface (Fig. 1.76). Two cervical and one thoracic vertebrae exhibit mild osteophytic lipping. The left fourth metatarsal and its associated proximal phalanx do not retain their bony articulation; the articular ends of these bones are severely eroded (Fig. 1.77).

Bone loss pathologies include notching of the acromion processes, and lesions on the articular surfaces of seven carpals. Alveolar erosion is associated with the maxillary canines and interproximal caries are evident between the upper right first premolar and canine, and the upper left canine and second premolar. The upper left first premolar was lost antemortem (Fig. 1.78).

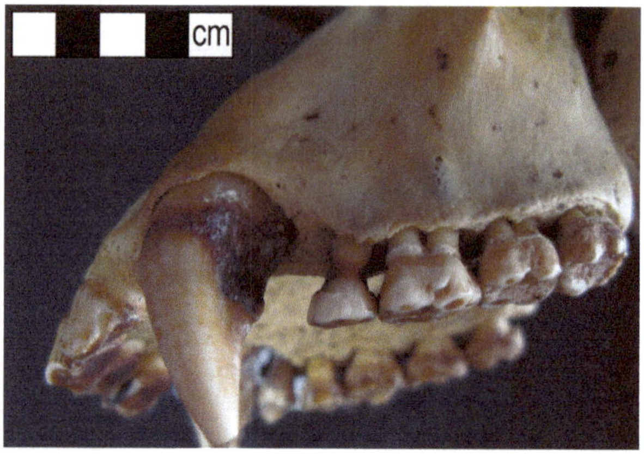

Fig. 1.78 Lateral view of Atlas's left maxilla, exhibiting antemortem loss of the first premolar

Fig. 1.79 Atlas's atlas

Two proximal and one intermediate manual phalanges have asymmetrical palmar muscle attachment sites, and two manual intermediate phalanges show proliferative bone consistent with periosteal reaction. The first cervical vertebra has asymmetrical transverse foramina (Fig. 1.79).

Beethoven (Cranium Only: Identity Tentative) (Male: 33.5 Years)
The recovery of this cranium from the Kasekela community's range coincided with the disappearance of adult male Beethoven in November 2002 (GSRC unpublished data; Williams et al. 2008). Beethoven was first observed in the Kasekela community in December 1973, when he is estimated to have been 4.5 years old (GSRC).

His unhabituated mother disappeared around the same time, orphaning Beethoven. This resulted in delayed growth and sexual maturation for him. Beethoven was "adopted" by a suspected older female sibling (estimated to have been about 9 years old at the time). He also associated closely with Jomeo (Goodall 1986), an adult male who is likewise part of the Gombe skeletal collection.

Beethoven is suspected to have succumbed to a gastrointestinal affliction, or "wasting disease" (Williams et al. 2008). The cranium shows few signs of trauma and pathology. Only the third molars right second molar, and left first premolar are in situ, though most of the remaining alveoli are consistent with the adult dentition being in full occlusion during life. Resorption of the alveoli for the left central incisor and left first molar indicate antemortem loss of these teeth (Fig. 1.80).

Echo (Female: 34 Years)

Echo was first observed in the Kasekela community in February 2004, when she is estimated to have been 19.6 years old. She is hypothesized to be from the unhabituated Kalande community, as she was observed during an intercommunity encounter in September 2003. At the time Echo emigrated to Kasekela, she brought her daughter, Eowyn, with her, then estimated to have been approximately 3 years old. The adult female Eliza (age 21.4 years in November of 2016), a current member of the Kasekela community, is hypothesized to be Echo's firstborn. Another daughter, Emela, was born to Echo in October 2005, but likely did not survive her mother's death in 2006 (GSRC).

In 2006, Echo died of complications from a massive spinal cord injury (T5) described elsewhere (Terio et al. 2011 (Echo is referred to as Ch-099 in this publication)). A section of the thoracic vertebral column was removed *en bloc* during necropsy for analysis and as such is not included in the skeletal analysis (it is a preserved soft tissue specimen). Echo was unable to use her lower extremities at all during the last period of her life and dragged herself around using only her forelimbs. Nutritional deficit may have been a contributing cause of death, and constant abrasion with the forest substrate may have contributed to the development of abdominal abscesses described by Terio et al. (2011). There are four cervical and

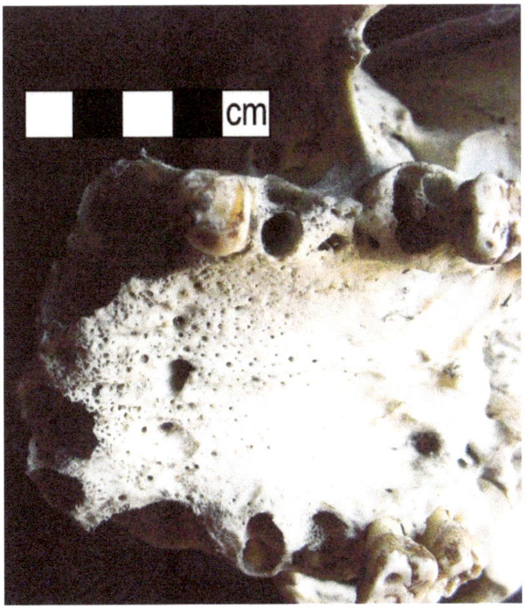

Fig. 1.80 Beethoven's maxillary dentition

nine thoracic vertebrae associated with her dry skeletal remains, and the skeleton is otherwise remarkably complete. The sternum and hyoid are not preserved, and she lacks one phalanx (55 total are associated with her skeleton). All other skeletal elements are present. Most of the epiphyses are fully fused, though the medial clavicle is less than half fused and the medial scapular margin and the iliac crest still show prominent epiphyseal lines. The complete adult dentition is in full occlusion.

Several traumata in addition to the spinal column injury are evident. These include a well-healed fracture to a manual proximal phalanx, and osteophytic growth associated with palmar muscle attachment sites on two metapodials and a pedal proximal phalanx, possibly associated with trauma to these muscles. Eight of the ribs show an abnormal contour, suggesting possible well-healed fractures, though no bone callus is evident.

The left and right tibiae as well as an intermediate manual phalanx show signs of reactive bone on the periosteal surface, a generalized sign of inflammation. The right talus has smooth, osteophyte-like growths unassociated with the articular surface (Fig. 1.81). The bicipital groove of the left humerus has "bridging" osteophytes indicative of partial ossification of the transverse humeral ligament (Fig. 1.82). The right clavicle has osteophytes on its acromial articulation, and the superior angle of

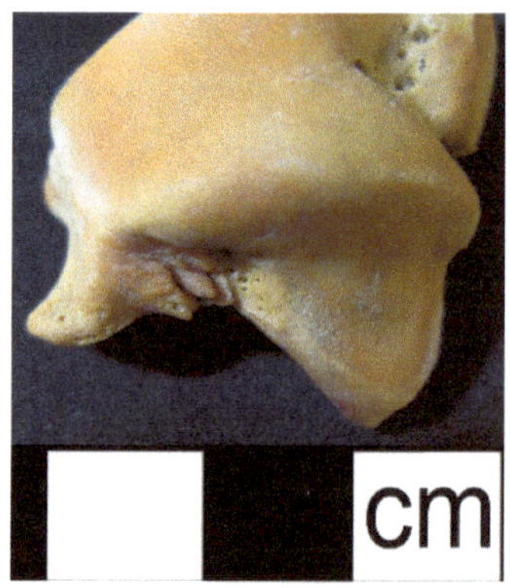

Fig. 1.81 Echo's right talus

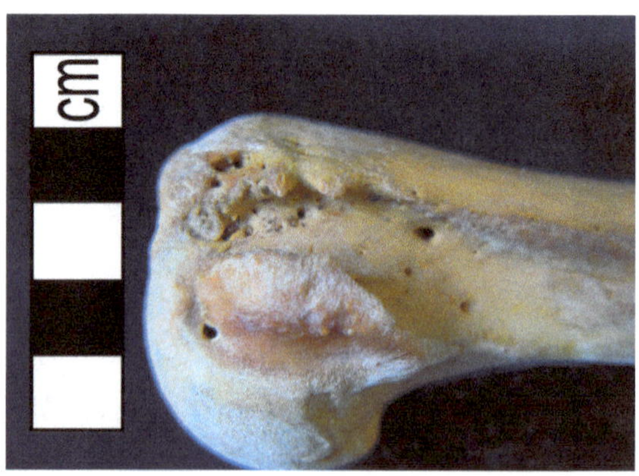

Fig. 1.82 Partial ossification of Echo's transverse humeral ligament

the right scapula also exhibits bone growth that is osteophytic in appearance (Fig. 1.83). The right olecranon process displays an area of proliferative bone, though this lesion is diffuse. These particular bone formation pathologies may be related to the forelimb-reliant locomotion upon which Echo was forced to depend.

The most dramatic of the proliferative lesions occur on the pubic bones, and are likely to have been associated with the abdominal abscesses described by Terio and colleagues. The right pubis in particular shows severe bone involvement, the lesion being more than 2 cm in diameter with an extensive fistula (Fig. 1.84).

Congenital variations include unusually long transverse processes on the first lumbar vertebra (Fig. 1.85), and laminar deficiency in the first through fifth sacral vertebrae (Fig. 1.86).

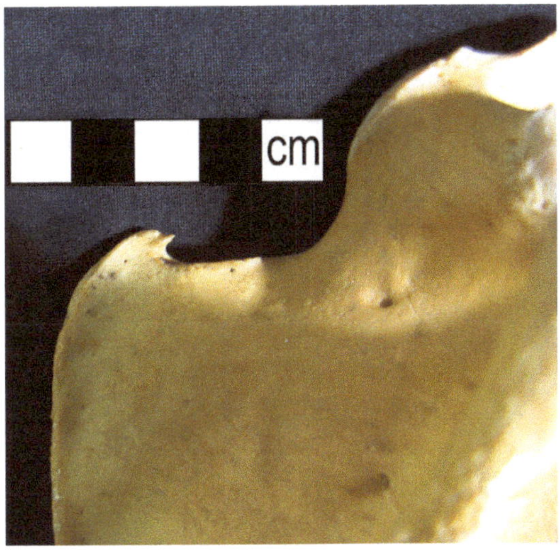

Fig. 1.83 Osteophytosis on the superior angle of Echo's scapula, possibly indicating overuse of levator scapulae

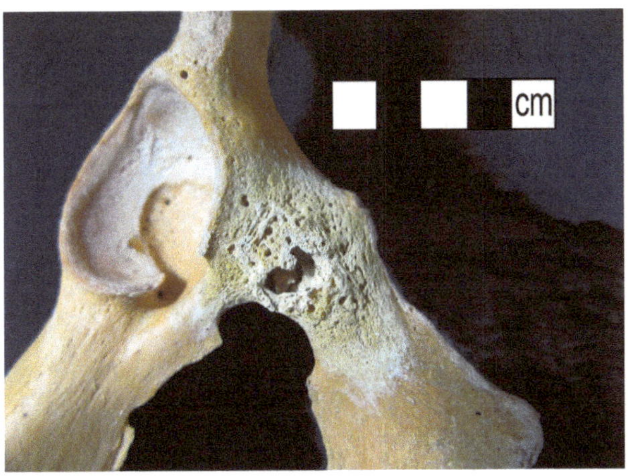

Fig. 1.84 Antero-lateral view of Echo's right os coxa

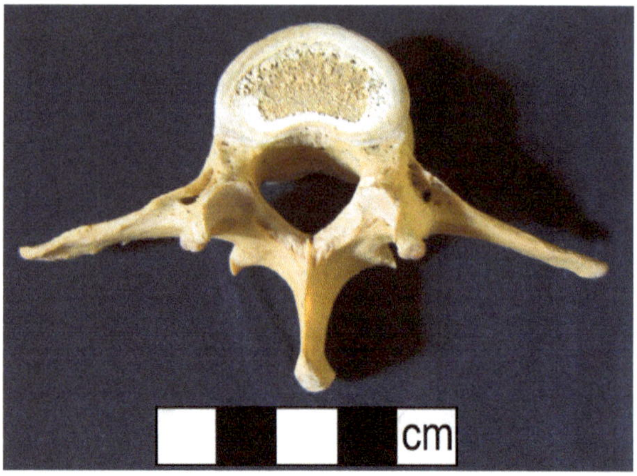

Fig. 1.85 Unusually long transverse processes on Echo's first lumbar vertebra

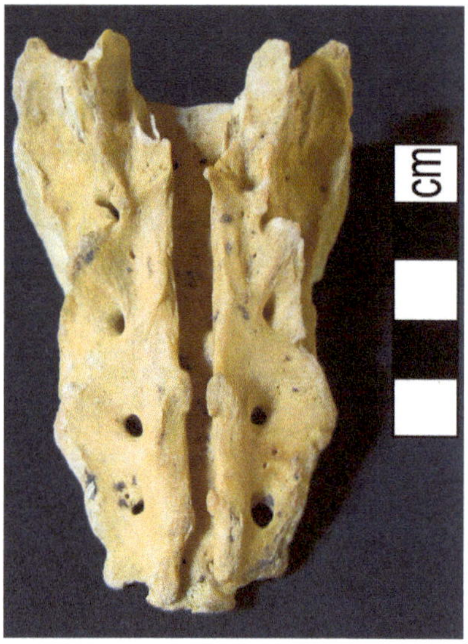

Fig. 1.86 Laminar deficiency in Echo's sacrum

Humphrey (Male: 34.9 Years)

Humphrey was a very aggressive alpha male when in his prime (Goodall 1986). He is hypothesized to be Melissa's sibling because of their close association patterns (Ibid.) (Melissa is also part of the Gombe skeletal collection, see below). Humphrey was first observed in the Kasekela community in July of 1963, when he is estimated to have been 17 years old. Humphrey died in June 1981 of unknown causes (Williams et al. 2008).

Only the cranium is preserved. Most of the dentition is moderately worn (Fig. 1.87), though the canines have exposed pulp cavities and were likely broken antemortem. There are periapical abscess drainage sites associated with both canines as well as the right central incisor (Kilgore 1989).

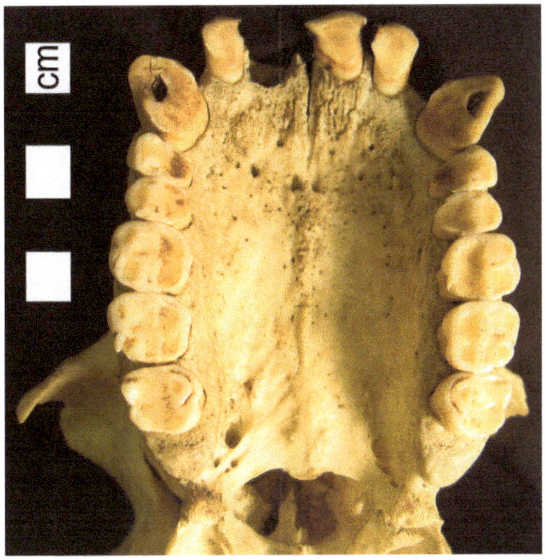

Fig. 1.87 Humphrey's maxillary dentition. The right central incisor was lost postmortem

Melissa (Female: 36.3 Years)

A high-ranking Kasekela female, Melissa died in 1986 of a "wasting disease" (Williams et al. 2008). She was first observed in August 1963, when she is estimated to have been 13 years old. Her first recorded pregnancy was Goblin, born in September 1964; he eventually became the alpha male of the Kasekela community and is also part of the Gombe chimpanzee skeletal collection (Goodall 1986; GSRC). His case is discussed below.

Melissa had eight recorded pregnancies. After Goblin, she gave birth to a female infant in 1969, who disappeared after just a few days. Her daughter, Gremlin, was born in 1970 (Goodall 1986), and is currently the oldest female in the Kasekela community; she will be 49 years old in 2019 (GSRC). Melissa gave birth to another male infant in 1976, who is suspected to have been killed by Passion and Pom in his first week of life (see first case in this chapter). Genie, a female born in 1976, was observed to be killed and eaten by Passion and Pom before she was a month old (Goodall 1986).

In October 1977, Melissa gave birth to male twins, Gimble and Gyre. Gyre died of a respiratory infection before he was a year old (Ibid.). His skeleton is part of the Gombe collection, and his case is described above. Gimble, however, survived to adulthood, staying in the Kesekela community until his disappearance in January 2007.

Interestingly, both Melissa's daughter Gremlin and Gremlin's daughter, Gaia, have also birthed twins (GSRC). Melissa's last offspring, Groucho, was born in July 1985, and died shortly before his mother of a gastrointestinal disease in October 1986 (Goodall 1986; GSRC). Groucho is also included in the skeletal collection, and his case is discussed above.

Melissa was affected by the polio epidemic, and though she regained the use of her arms, her head remained permanently crooked to the left (Goodall 1986), suggesting torticollis caused by hypertension or scarring of the right sternocleidomastoid muscle. While deinnervation of the sternocleidomastoid in humans does not usually result in an abnormal head position, it may be possible that Melissa's spinal accessory nerve was affected by her bout with polio.

Her skeleton is nearly complete; her coccyx is not preserved, and 30 carpals/tarsals, 25 ribs, one patella, and 47 total phalanges are associated with her otherwise complete skeleton. Her epiphyses are all fully fused, except for the left acromion process, which is still patent, though more than half fused (Fig. 1.88). This is likely due to repetitive motion rather than lack of skeletal maturation (see notes on this phenomenon in the section about Pallas), and it is probably significant that her left acromion is thus affected. Given the left lateral flexed position of her neck after surviving her bout with polio, Melissa may not have fully regained the use of all the muscles of her neck/upper limb on the right side. Her skeleton, however, does not demonstrate any marked size asymmetry like other polio survivors such as Madam Bee and Gilka, likely because Melissa was already an adult by the time of the outbreak (Morbeck 1987). This sign of significant repetitive motion is also evident on several other animals unaffected by the polio epidemic (e.g., Pallas, Flo), and Melissa's incompletely fused acromion may therefore be unrelated to her infection with or recovery from polio.

An epiphyseal line is still visible on her proximal humerus, proximal radius, femoral head, and proximal fibula. The complete adult dentition is in full occlusion, and exhibits moderate to heavy wear (Jurmain and Kilgore 1995; Kilgore 1989). The left maxillary lateral incisor was likely broken during life. The pulp cavity is exposed and the occlusal surface of the tooth exhibits wear (Fig. 1.89), though there is no associated abscess drainage site.

The left first metacarpal has an angular deformity consistent with a well-healed fracture. Another partially-healed fracture is evident on an intermediate manual phalanx. Uneven plantar muscle attachments on a proximal pedal phalanx suggest a muscle injury resulting in bony remodeling. This phalanx and an additional pedal phalanx exhibit bony calluses that may be evidence for well-healed fractures. One manual intermediate phalanx exhibits reactive bone indicative of periosteal reaction. Reactive bone on the right transverse processes of the fourth lumbar and first sacral vertebrae indicate that these elements were beginning to fuse (Fig. 1.90),

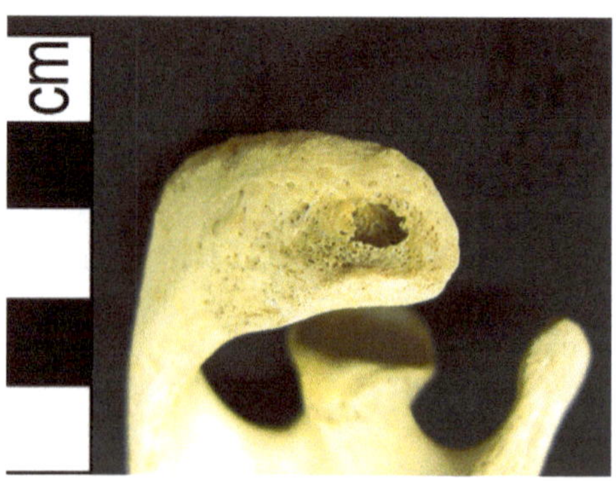

Fig. 1.88 Melissa's unfused acromion

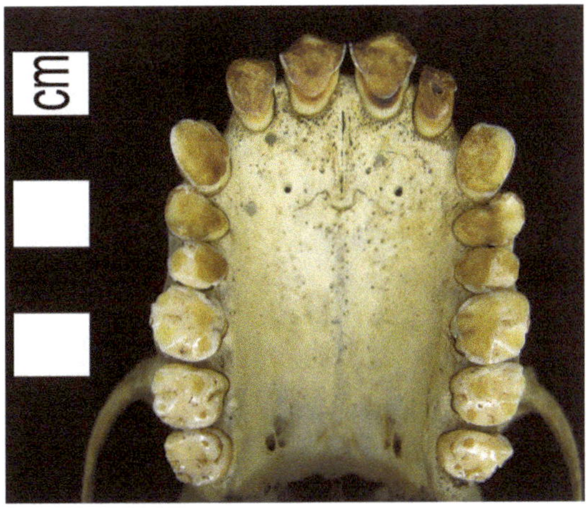

Fig. 1.89 Melissa's maxillary dentition

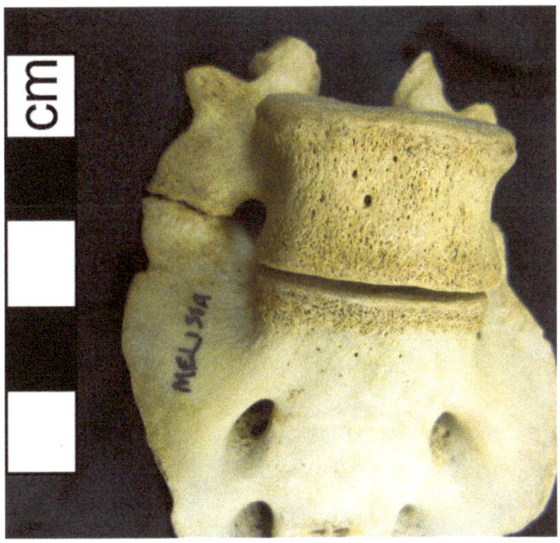

Fig. 1.90 Incipient sacralization of Melissa's L4

and medially-extending osteophytes on both auricular surfaces indicate that the sacroiliac joints were beginning to fuse. In addition, several unprovenanced bony fragments are curated with Melissa's skeleton (Fig. 1.91). If they were associated with her body, they may be evidence for heterotopic ossification, possibly subsequent to a serious injury (Barfield et al. 2017).

Hugo (Male: 38.6 Years)
A high-ranking male known for his "leadership abilities" (Goodall 1986, p. 69), Hugo died in 1975 of a respiratory infection (Williams et al. 2008; Ibid.), in the same month as his close companion Mike (Goodall 1986). He was the fourth adult male to visit Goodall's research camp in May of 1963 (Ibid.), when he is estimated to have been 27 years old (GSRC).

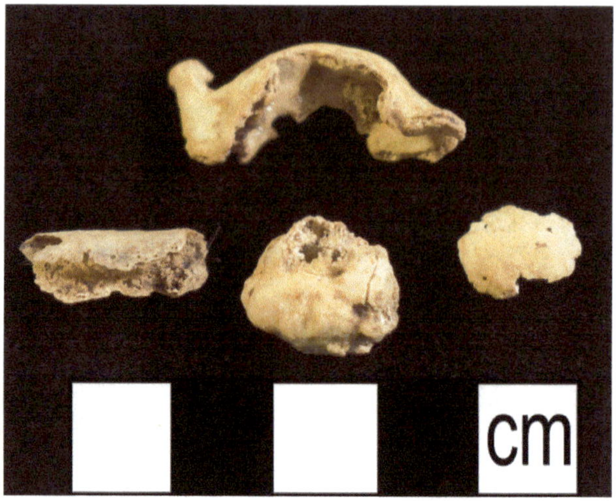

Fig. 1.91 Evidence for heterotopic ossification from Melissa's skeleton

Hugo is represented by a partial skeleton: both femora and the left os coxa are not preserved. The greater cornua, but not the body, of the hyoid are preserved. There are 25 carpals/tarsals and 28 total phalanges associated with the skeleton. All other elements are present and the epiphyses are all fully fused. Epiphyseal lines are still visible on the distal ulna and distal fibula.

All of Hugo's adult dentition had erupted at one point, but the mandibular posterior teeth were all lost antemortem. The remaining maxillary posterior dentition is moderately worn (Kilgore 1989). The maxillary first premolars are likely to have been lost shortly before death; the alveoli are patent but reactive bone is evident.

Jurmain (1989) described at least eight fractures on Hugo's skeleton, including to the right zygomatic arch (Fig. 1.92), the spinous process of the third cervical vertebra, the midshaft of two left ribs, the right calcaneus, the right fifth metacarpal, and two pedal phalanges. In addition, a callus on the distal end of the left 13th rib is likely another traumatic lesion.

Exposure of the pulp cavity in the dentition of wild animals, particularly long canines, may represent breakage of that tooth, especially if the rest of the dentition exhibits less wear. In Hugo's case, his remaining dentition is only moderately worn. His extensive antemortem tooth loss may be related to the same injury that caused his zygomatic arch fracture (as in Vincent's case). While his left maxillary canine was lost postmortem, the roots of the other canines remain, with exposed pulp cavities and evidence of wear on the (possibly) broken surfaces. There are periapical abscess drainage sites associated with both maxillary canines, and the right mandibular canine. An area of reactive bone in the periapical region of the left mandibular canine suggests an incipient fistula.

Many of Hugo's synovial joints exhibit evidence for osteoarthritis (osteophytes and/or eburnation), including: the proximal articulation of the left third metatarsal, left first and second cuneiforms, right calcaneus, right first cuneiform, the proximal

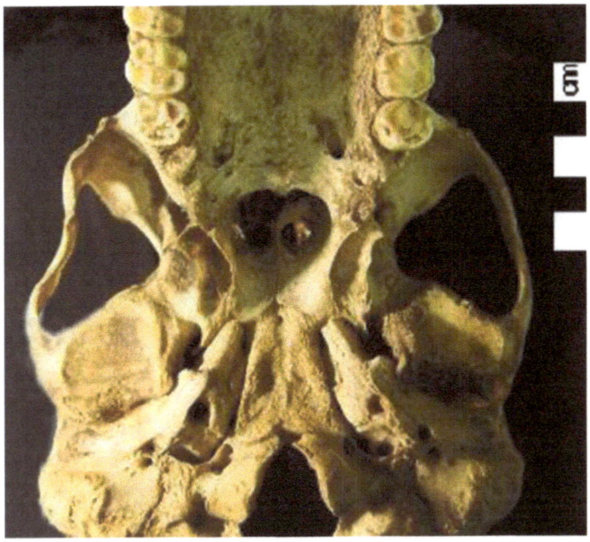

Fig. 1.92 Healed fracture to Hugo's right zygomatic arch

and distal articulations of both fibulae, the left proximal tibia, the left distal humerus, the proximal and distal articulations of both ulnae, the distal articulation of two metacarpals, and a manual proximal phalanx. His left mandibular fossa and mandibular condyle show (Fig. 1.93) evidence of a moderate presentation of osteoarthritis of the temporomandibular joint (*sensu* Rando and Waldron (2012)), with altered contour of the joint, increased porosity, bone formation, and eburnation. This case of osteoarthritis may be a sequel to the fracture to his zygomatic arch.

Hugo also exhibits reactive bone consistent with periosteal reaction on a pedal intermediate phalanx, six manual intermediate phalanges, and the left first and second metacarpals. Both of his humeri have a patent olecranon foramen (Fig. 1.94).

Cusano (Male: 39.9 Years)
Former alpha male of the Mitumba community, Cusano died in June 1996 at nearly 40 years of age, and his skeleton was not exhumed until August 2005 (GSRC). Postmortem damage to the skeleton is extensive. All of the bones have extremely friable surfaces and elements of the axial skeleton (ribs and vertebrae) are mostly fragmentary. Both scapulae are fragmentary, and only the right radius survives. Both femora survived, though the right is fragmentary, and only the left tibia is present. No humeri, ulnae, or fibulae were preserved. The cranium and mandible are some of the better-preserved elements, though the mandible exhibits a post-burial break near the symphysis. The adult dentition is in full occlusion, with no antemortem tooth loss, and all observable epiphyses are fused.

The right zygomatic arch has an irregular contour (Fig. 1.95) and the temporal-zygomatic suture is obliterated, suggesting a healed fracture on this side (the left temporal-zygomatic suture is still interdigitated). Possible tooth breakage may also be associated with this injury. The maxillary teeth with possible antemortem breakage

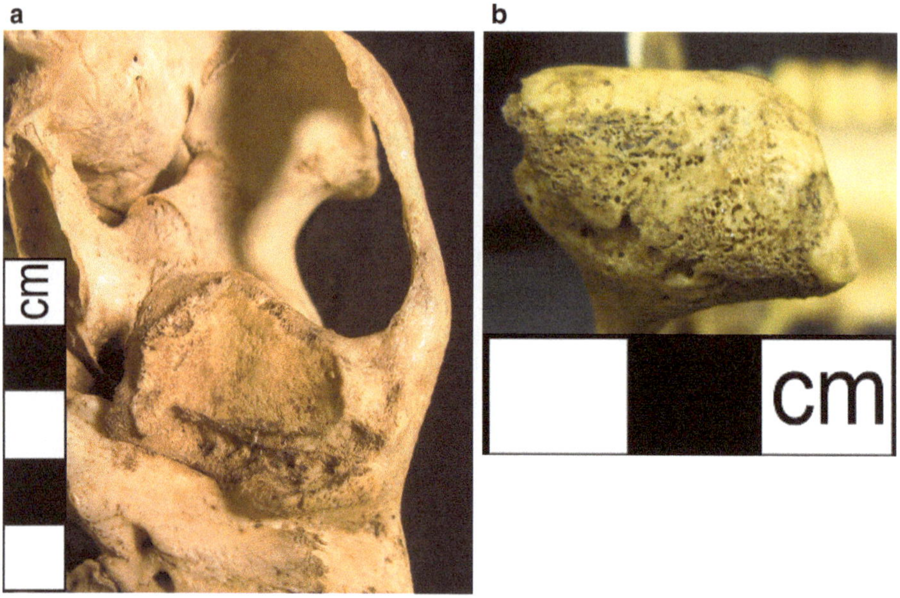

Fig. 1.93 (**a**) Hugo's left mandibular fossa, exhibiting bony changes consistent with osteoarthritis. (**b**) Hugo's left mandibular condyle, exhibiting osteoarthritis

Fig. 1.94 Patent olecranon foramen on Hugo's right humerus. The variation is present bilaterally

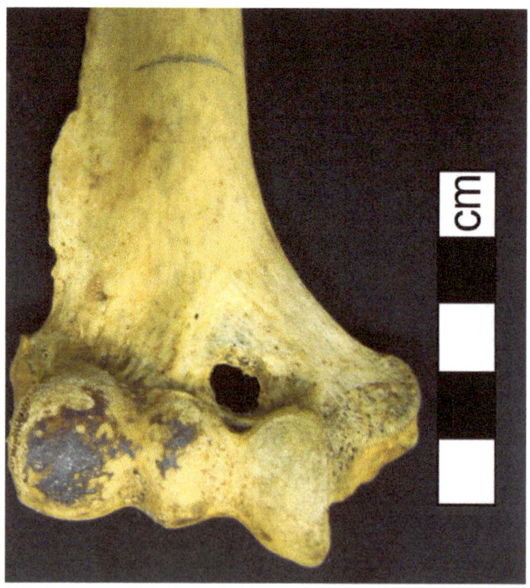

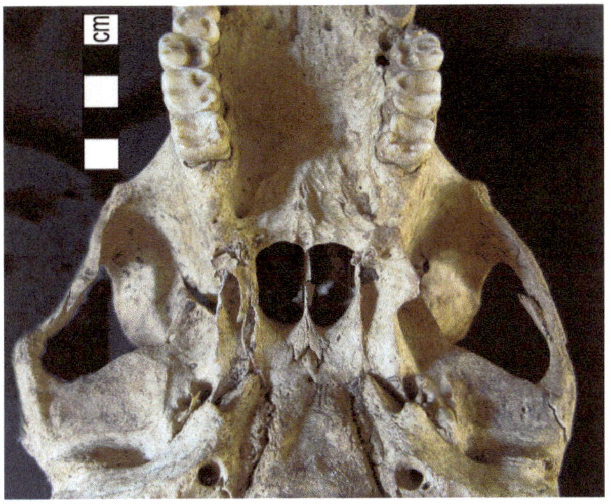

Fig. 1.95 Healed fracture to Cusano's right zygomatic arch. The left arch was broken postmortem

include: the right canine, right lateral incisor, left central incisor, left canine, and left first premolar. The left mandibular canine may have also been broken antemortem. The described damage to these teeth may be due to wear rather than breakage, but three of the four maxillary incisors are affected while the right central incisor is not, which supports the tooth breakage hypothesis in this case (adjacent teeth would be expected to wear at approximately the same rate). There are also five maxillary abscess fistulae (Fig. 1.96) ranging in size from approximately 1–4 mm in diameter, and the mandibular abscess associated with the left canine (Fig. 1.97). Interproximal caries are evident on the upper right first premolar, the upper right central incisor, the lower left first molar, and all of the mandibular incisors.

One cervical vertebra exhibits mild osteophytic lipping of the body, and the fourth lumbar vertebra is completely fused to the first sacral vertebra, though no osteophytic lipping is evidenced on either element and the disc space between the two is still evident. Due to his incomplete skeleton, Cusano is excluded from quantitative analyses on skeletal trauma and pathology, though his complete dentition allows for his inclusion in analyses of the dentition only.

Goblin (Male: 40 Years)

A former alpha male of the Kasekela community, Goblin was severely attacked by other chimpanzees twice. The first time was in November 1979, shortly after his initial rise to the alpha position. He sustained a severe laceration to his groin during a gang attack that occurred in association with a hunting episode. The wound bled for an hour. This injury restricted his mobility, and, starting 1 week after the injury, leaked pus for 2 weeks. Goblin's total recovery time from this injury was 5 weeks, after which he re-asserted himself in the dominance hierarchy and reclaimed the alpha position (Goodall 1986). Goblin was alpha male for approximately 10 years before being severely injured by Wilkie, the number two male, in September 1989. During their fight, both males fell out of a tree. Goblin bit Wilkie on the face, but in

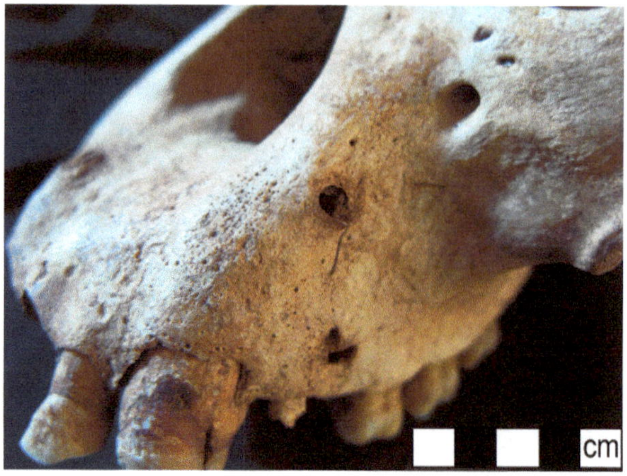

Fig. 1.96 Some of Cusano's maxillary abscesses, associated with the canine and premolars

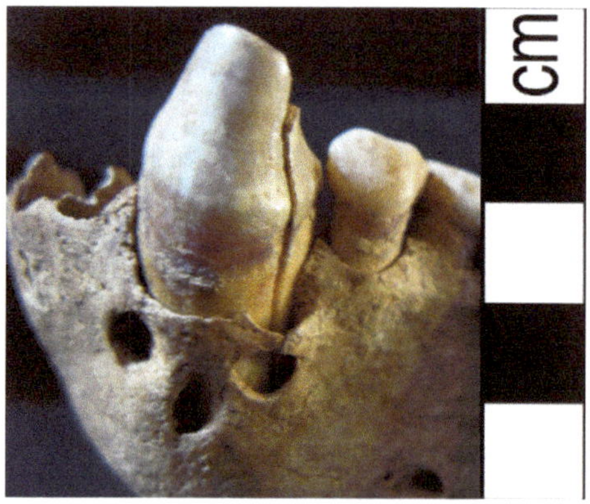

Fig. 1.97 Multiple fistulae for a periapical abscess associated with Cusano's left mandibular canine, which was likely broken antemortem, exposing the pulp cavity. Postmortem damage to the tooth (longitudinal crack) is also evident in this image

turn received bite wounds on his right wrist, left foot, several fingers and toes, and scrotum. The scrotal injuries were the most severe, and became infected. It is likely that Goblin's life was saved by human veterinary intervention in this case (Goodall 1992). Goblin finally succumbed to intestinal parasites many years later (Williams et al. 2008), just shy of his 40th birthday, on the 24th of August, 2004 (GSRC).

Goblin was first observed on the day of his birth: the 6th of September, 1964. He is Melissa's first recorded offspring, born when she is estimated to have been 14 years old. Goblin's case is fascinating because of the rich behavioral record available for this high-ranking male; records from Gombe include observations from the day of his birth, the day of his death, and many intervening days in his 40-year lifespan. In addition, many members of Goblin's family are also included in the Gombe skeletal collection, including his mother, Melissa. Also included in the skeletal collection are: Humphrey (Goblin's putative uncle), three maternal siblings (an unnamed male infant who died in 1976, Gyre, and Groucho), three nephews (Gremlin's unnamed infant who died in 1987, Getty, and Galahad), and a great-nephew (Gaia's male infant who died in 2008).

The skeleton is complete and well preserved, lacking only a few phalanges (49 are associated with his skeleton). The entire hyoid is present, and Goblin has 13 thoracic and five lumbar vertebrae (mode = 4 lumbar vertebrae). All secondary ossification centers and epiphyses are fully fused, with no lingering observable epiphyseal lines, and the sternal body is completely fused. The complete adult dentition was in full occlusion during Goblin's prime, though several dental pathologies of interest are discussed below.

Goblin exhibits relatively little evidence for healed trauma, consistent with the fact that his known major injuries primarily affected soft tissues. The first lumbar vertebra is missing the left transverse process, and the reactive bone at this site suggests this is a traumatic loss rather than a congenital variation. Well-healed fractures with relatively smooth calluses are evident on the left fifth metatarsal, eight ribs (four pairs), an intermediate manual phalanx, and an intermediate pedal phalanx.

The knees, ankles, hips, temporal-mandibular joints (both the temporal and mandibular components), three pedal phalanges and two sesamoids exhibit increased porosity and osteophytic lipping consistent with osteoarthritis; three pedal phalanges exhibit eburnation. All of the cervical vertebral bodies show mild to moderate osteophytic lipping, as do nine of 13 thoracic vertebrae, and three lumbar vertebrae. The fifth lumbar and first sacral vertebrae are fused.

The teeth are well worn, with dentin exposure over the entire occlusal surface (Fig. 1.98). The upper left first premolar was lost antemortem. Interproximal caries are evident on the anterior maxillary dentition as well as the right first premolar, the mandibular premolars and incisors. The mental foramina are both enlarged and show osteophytic growth (Fig. 1.99). The maxillary canines and central incisors also have associated abscess drainage sites (Fig. 1.100). The anterior maxillary dentition has advanced pulp cavity exposure, especially for the left canine, whose

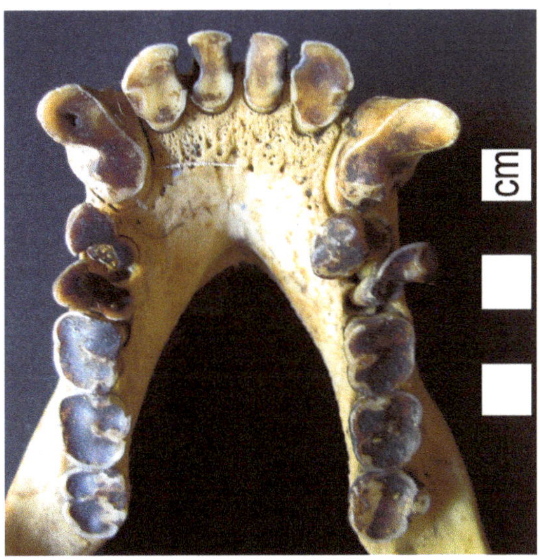

Fig. 1.98 Goblin's mandibular dentition, showing extensive dentin exposure as well as several dental pathologies (see text)

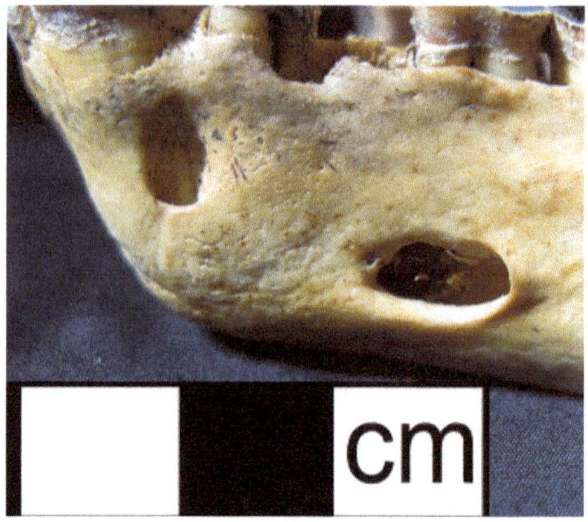

Fig. 1.99 Goblin's enlarged mental foramen, exhibiting reactive bone internally (left side pictured; lesion is present bilaterally). Note also the fistula for a periapical abscess likely associated with the canine

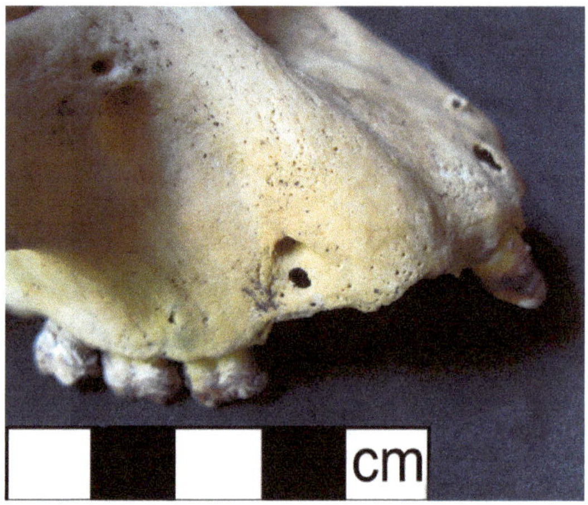

Fig. 1.100 Some abscess drainage sites associated with Goblin's maxillary dentition

alveolus is severely eroded. The crown of the left maxillary canine is entirely gone (Fig. 1.101). Pulp cavity exposure is much more limited on the anterior mandibular dentition, involving only a small exposure at the tip of the left canine. This suggests the possibility that the maxillary dentition may have been broken rather than simply worn.

Osteophytic growths partially bridging the bicipital groove indicate partial ossification of the transverse humeral ligament. Lastly, some unidentified bony fragments were recovered with Goblin's skeleton and are likely the abnormal ossification of soft tissue (Fig. 1.102). The irregular morphology of these fragments is consistent with heterotopic ossification, which is a common sequel to amputation, massive soft tissue trauma, and head trauma/traumatic brain injury in humans (Barfield et al. 2017). While the exact provenance of these fragments is, unfortunately, not known, it may be that they represent heterotopic ossification or a similar disease process in Goblin, subsequent to one or more of the severe injuries he sustained during life.

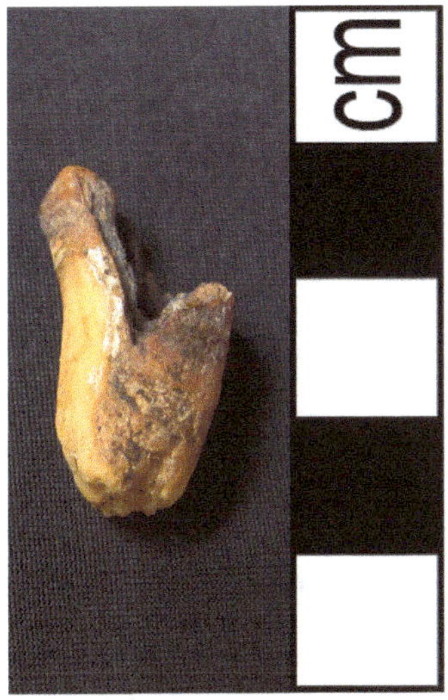

Fig. 1.101 Goblin's left maxillary canine was still in situ at the time of his death, but breakage and pulp cavity exposure contributed to severe periodontal disease associated with this tooth

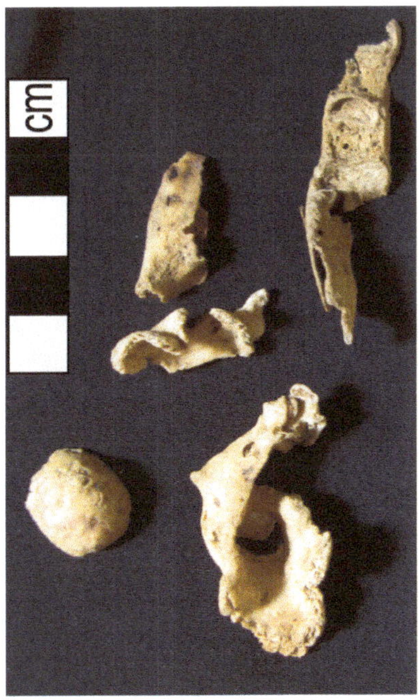

Fig. 1.102 Unidentified bone fragments associated with Goblin's skeleton, which may have resulted from heterotopic ossification

Patti (Female: 44.3 Years)

Patti was first observed in the Kasekela community in March 1971, when she is estimated to have been nearly 10 years old. Her first recorded pregnancy, a male infant who died in 1978 at approximately 1 week old, is also part of the Gombe skeletal collection; his case is described above. Patti's subsequent pregnancies include a male infant born in 1979, Tapit, who succumbed to an unknown illness at

5 years of age, Tita, a daughter born in 1984 who disappeared or emigrated away from Gombe at 11 years old, and an infant of unknown sex born in 1988 who was never extensively observed and died within a week of birth. Patti's daughter Tanga, born in 1989, remains in the Kasekela community, and turned 30 years old in April 2019. Patti's son Titan, born in 1994, disappeared in March 2015, after observers noticed he was ill. He was an impressive, prime-aged male at the time of his disappearance, poised to take over as alpha male. Patti's last pregnancy was a son, Tarzan, born in 1999. He was 6 years old at the time of his mother's death, after which he associated more closely with Tanga (his older sister). Tarzan died in 2013, age 14, of an unknown illness (GSRC).

Patti was attacked by male chimpanzees from the Mitumba community in October 2005, and died because of her injuries 13 days later (Williams et al. 2008). Her skeleton is complete and well preserved; only the hyoid and one phalanx were not recovered. Interestingly, Patti has 12 thoracic vertebrae (mode = 13) with 24 paired ribs, and five lumbar vertebrae (mode = 4). Secondary ossification centers are all fused. The fully fused medial margin of the scapula still has a visible epiphyseal line. The complete adult dentition was in full occlusion during Patti's prime.

Few examples of trauma are observable on Patti's skeleton. A puncture in the left calcaneus (Fig. 1.103) is the only example of perimortem trauma. Impression at necropsy was that extensive soft tissue trauma had occurred (K. Terio unpublished data). Well-healed fractures on two ribs and a manual intermediate phalanx are evident. The maxillary right second premolar is chipped, and wear on the broken surface suggests the break occurred antemortem.

The dentition is overall in excellent condition considering Pati's age, with only moderate occlusal wear. The only tooth lost antemortem was the upper left first molar (Fig. 1.104). The mandible exhibits moderate periodontal disease: both mental foramina are enlarged and alveolar erosion is evident in association with the posterior dentition on the left side. Both ulnae and one rib exhibit reactive bone

Fig. 1.103 Perimortem puncture wound in Patti's left calcaneus

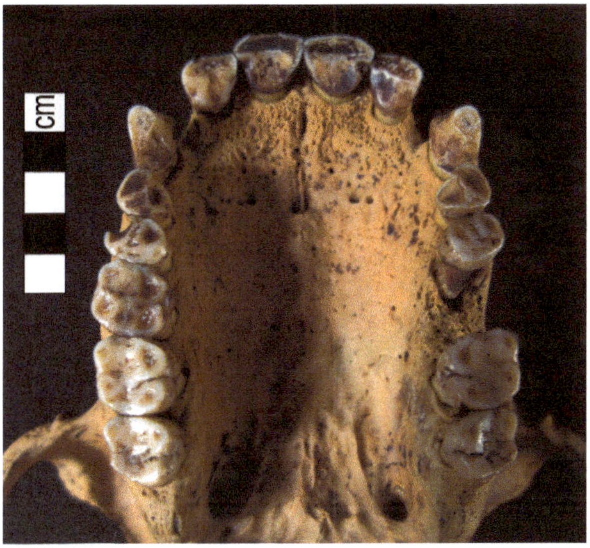

Fig. 1.104 Patti's maxillary dentition, exhibiting only moderate wear as well as the antemortem loss of the upper left first molar

consistent with mild periostitis. Another rib has a smooth-looking callus on its anterior aspect; this lesion may not be a result of trauma, as the rest of the rib's contour is regular (Fig. 1.105). Eburnation of the joint surface is observable on the distal femora (Fig. 1.106), proximal and distal ulnae, two manual intermediate phalanges, and three metatarsals. In addition, seven vertebrae are affected by osteophytic lipping.

Bwavi Male (Unknown) (40 Years?)
This unknown male from the Kalande community wandered into the Kasekela range towards the end of his life in 1994 or 1995. Field observers remember this particular individual (though not the exact year his case was noticed). He was clearly ill, did not eat, and died soon after first being observed (GSRC unpublished data). His teeth are not as worn as Goblin's, but are more worn than Patti's (especially the mandibular first molars), and his skeleton displays a remarkable number of pathologies, particularly severe joint disease.

His skeleton is exceptionally complete, including all 56 phalanges, the entire hyoid and one coccygeal vertebra. All secondary ossification centers are fully fused, though epiphyseal lines are still visible on the sacrum. The complete adult dentition is in full occlusion and no antemortem tooth loss is evident. The mandibular first molars, however, are extremely worn, with most of the crown eliminated. The roots alone remain on the buccal aspect. Alveolar pocketing is associated with both of these teeth (Fig. 1.107).

All of the joint surfaces in the skeleton, as well as the vertebral bodies, are severely affected by degenerative joint disease. Significant remodeling of articular surfaces is common, as well as extensive pitting, eburnation, and osteophytic growth at the margins of articular surfaces (Fig. 1.108). The case is particularly interesting

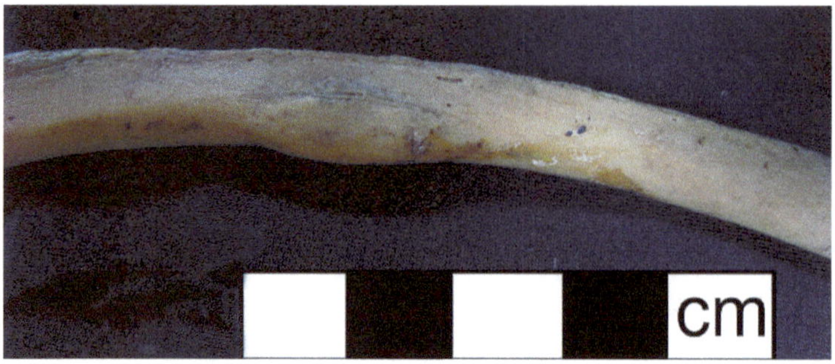

Fig. 1.105 Bone formation pathology on Patti's rib

Fig. 1.106 Degenerative joint disease on Patti's distal femur

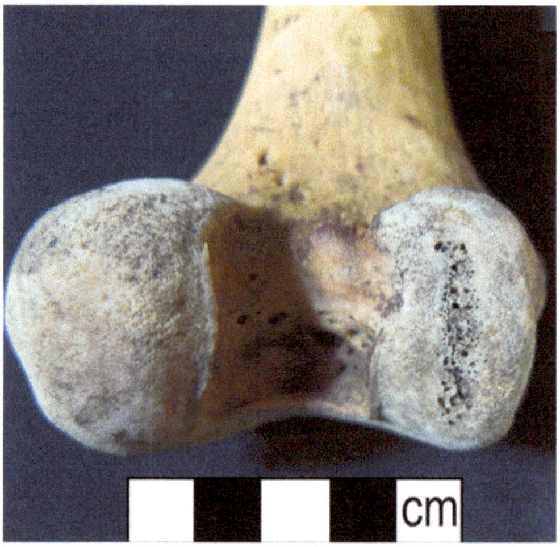

Fig. 1.107 Bwavi male's mandibular dentition, showing severe wear on the first molar (wear is bilateral), but only moderate wear on the other teeth

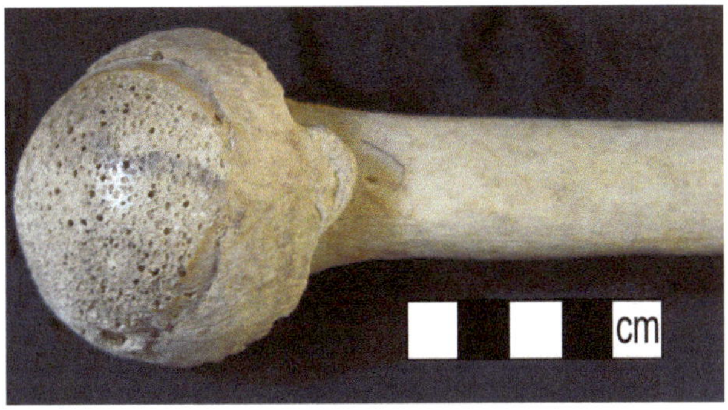

Fig. 1.108 Extensive remodeling and eburnation of Bwavi male's humeral head

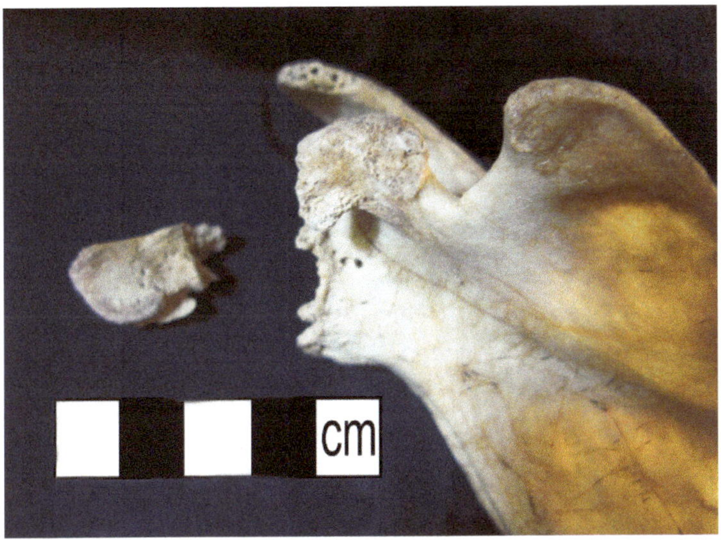

Fig. 1.109 Non-union fracture to Bwavi male's right coracoid process

because degenerative joint disease seems less common in African apes compared to humans (Jurmain 2000), and because there is no clear skeletal indicator of why this animal was so severely affected. While joint degeneration can be a sequel to trauma (Ibid.), the skeletal traumata evident on this animal are in no way unusual for an older male chimpanzee.

Evidence for skeletal injuries includes well-healed fractures on six ribs and both of the fifth metatarsals. A partially-healed, non-union fracture of the right coracoid process is evident (Fig. 1.109). The left third and fourth metacarpals are fused together proximally, possibly as a result of healing trauma (Fig. 1.110), as are the intermediate and distal pedal phalanges of a single ray. In addition, both femora exhibit osteophytic growth on the trochanters, as does the left ischial tuberosity.

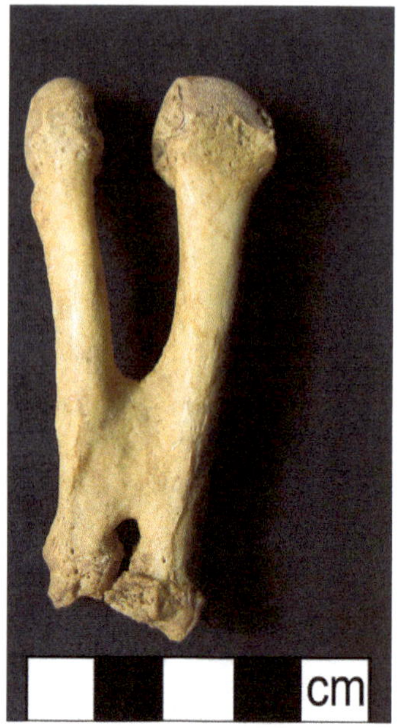

Fig. 1.110 Bwavi male's left third and fourth metacarpals, fused near their bases

Old Female (50 Years Plus)

The body of an elderly female chimpanzee was discovered in August 1973 by Y. Selemani (Richard Wrangham, personal communication; Goodall 1986). She had probably been dead at least one day. Community membership is attributed to the unhabituated Kalande community as she was not an identified individual (Wrangham et al. 2006). Wounds on her body and her body's position are consistent with intra-specific violence as a cause of death. Her body was discovered when males from the Kahama community made a convoluted detour in order to find it. Their knowledge of the body's location implies that they had prior knowledge of the location and had likely been involved in the attack (Richard Wrangham, personal communication; Goodall 1986).

Her skeleton is mostly complete, but poorly preserved; bone surfaces are very friable, especially hand and foot bones. The left ulna is preserved, but not the right. There are 25 carpals/tarsals, 45 total phalanges, five cervical vertebrae, and 25 ribs associated with this skeleton, and no coccyx or hyoid. The degree of dental attrition is similar to Flo's; they may have been of similar age at the time of death.

A non-union fracture to the right mandibular corpus was still in the process of callus formation at the time of death. A well-healed, but severe, fracture to the distal shaft of the right humerus is also evident. The right humerus was distinctly shorter than, but also broader than, the left (Jurmain 1989). This animal also exhibits well-healed fractures to the spinous process of L4, and one pedal intermediate phalanx. A pedal proximal phalanx has a non-union fracture that was in the process of callus formation at the time of death. She also exhibits a bone formation pathology on the right femur, superior to the lateral condyle (Fig. 1.111), which may have resulted

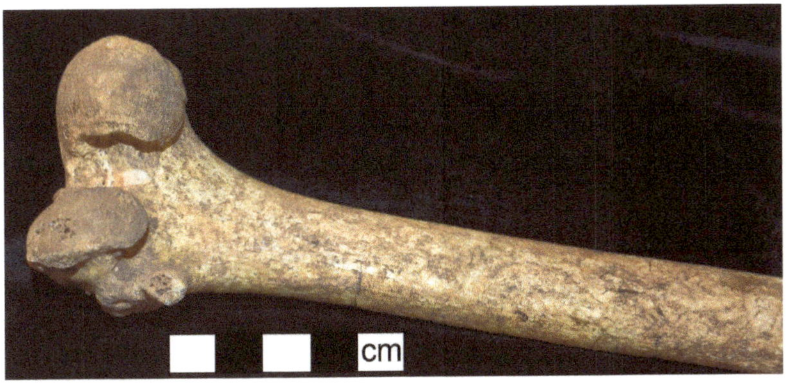

Fig. 1.111 Abnormal bone growth on Old Female's lateral epicondyle

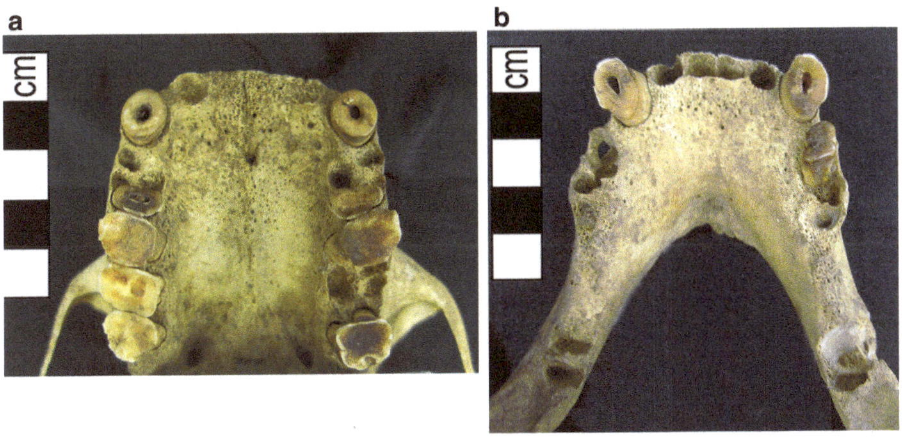

Fig. 1.112 (**a**) Old Female's maxillary dentition, exhibiting heavy wear. (**b**) Old Female's mandibular dentition

from an injury involving the gastrocnemius muscle, the lateral head of which originates at this location.

Her teeth are severely worn (Kilgore 1989). Several teeth are worn to the root and/or were broken, so that the pulp cavity is exposed: the right maxillary second premolar and all the canines. Several patent alveoli show evidence for reactive bone that indicates these teeth may have been lost near the time of death, or might have been shed soon, had Old Female lived longer: the right maxillary first premolar, the left maxillary first premolar, the left maxillary third molar, both left mandibular premolars, and the right mandibular third molar. Teeth lost antemortem, based on closed or drastically remodeled alveoli include: all maxillary incisors, the left maxillary second molar, all three right mandibular molars, the right mandibular second premolar, the right mandibular first molar, and the right mandibular second molar (Fig. 1.112). Drainage sites for periapical abscesses are evident near both maxillary canines, the right maxillary lateral incisor, and the left mandibular first premolar.

Eburnation is apparent on both distal femora, both glenoid fossae, the left proximal ulna, the right distal humerus, the proximal end of a metatarsal, the left acetabulum, and the right proximal tibia.

Flo (Female: 53.1 Years—Possibly Older)
Flo was a high-ranking, aggressive female from the Kasekela community when she was first observed by researches in May 1963. She is estimated to have been 44 years old at the time, and regularly traveled with two of her offspring, Figan, who is estimated to have been 10 years old in 1963, and Fifi, then aged 5 years. Because of her interactions with the adult male Faben, estimated to have been 16 years in 1963, he is also hypothesized to be Flo's son. Given that Flo would have been 28 when Faben was born (Goodall 1986; GSRC), he is unlikely to have been her first pregnancy.

Flo gave birth to a male infant in March 1964, Flint, who is also included in the Gombe chimpanzee skeletal collection (see above), as is Flo's grandson, Fred. Flo's last pregnancy, a male infant named Flame, disappeared in February 1969 during an illness so severe, Flo was not even able to climb a tree at night (Goodall 1986). Flo died of a gastrointestinal disorder (Williams et al. 2008) in August 1972 (Goodall 1986; GSRC), after which an obituary for her was published in the *Sunday Times* in the UK (Goodall 1986). Flint died a month after his mother, but Fifi went on to become one of the most reproductively successful females from Gombe of all time (Pusey et al. 1997), giving birth to nine offspring, seven of whom survived past infancy (GSRC).

Flo's skeleton is complete, lacking only a few hand and foot bones, the hyoid, and two ribs. There are 23 carpals/tarsals, eight metatarsals, eight metacarpals, 50 phalanges, and 24 ribs (for 13 thoracic vertebrae) associated with her skeleton. Most of the epiphyses are fused, though the acromion processes are not (see notes in section on Pallas for further discussion), and the medial margin of the scapula and iliac crest are more than half fused, but a gap is still evident between the primary and secondary ossification centers.

Flo's teeth are severely worn, and several were lost antemortem (Kilgore 1989), including all of the maxillary incisors, the left mandibular first and second molars, the mandibular lateral incisors, the right mandibular second premolar, and the right mandibular first and second molars. In addition, the alveolus for the right mandibular first premolar is so shallow and remodeled it was likely shed not long before death. Teeth with an exposed pulp cavity include the right maxillary first molar, all four canines, and the left maxillary first premolar. Fistulae for periapical abscesses are associated with the right maxillary first molar (palatal aspect), both maxillary canines (labial aspect), the left maxillary first molar (buccal aspect), and the right mandibular canine (labial aspect). Figure 1.113 exhibits Flo's dentition. A large osteophytic growth on the left coronoid process (Fig. 1.114) may be an indicator of altered, pathological masticatory biomechanics that developed as a result of Flo's numerous dental pathologies and/or facial trauma (see below).

Healed fractures are evident in the right clavicle, distal shaft of the right ulna (Fig. 1.115), and both fifth metatarsals (Jurmain 1989), as well as a metacarpal, two manual phalanges, one pedal phalanx, and the left zygomatic arch. The palmar

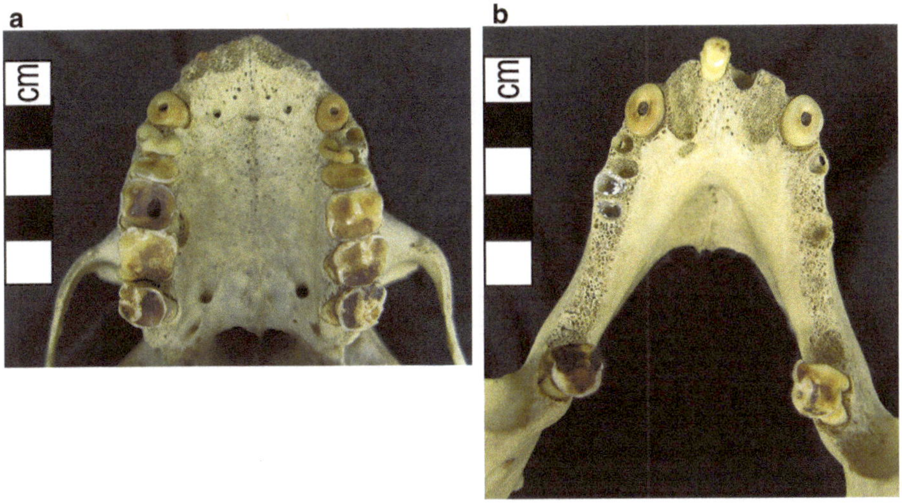

Fig. 1.113 (**a**) Flo's maxillary dentition. (**b**) Flo's mandibular dentition

Fig. 1.114 Osteophytic growth on Flo's left coronoid process

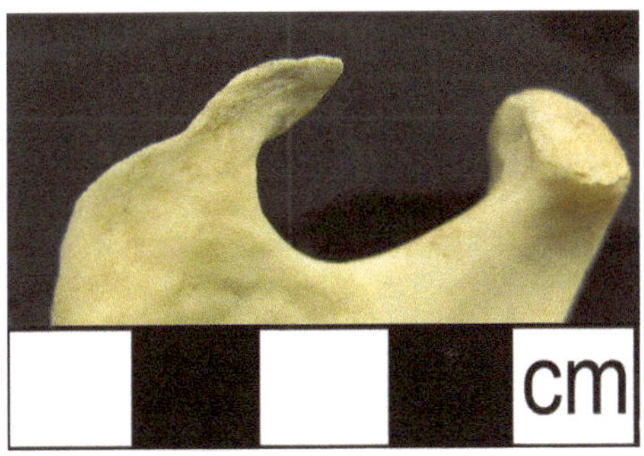

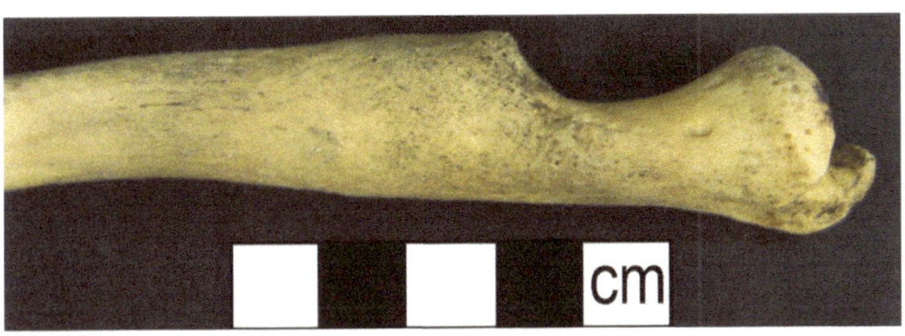

Fig. 1.115 Healed fracture to Flo's right ulna

muscle attachments on one manual intermediate phalanx are uneven, as are those on two pedal proximal phalanges, and the right fifth metatarsal and an additional metacarpal exhibit periosteal reactive bone. Bone growth pathologies are evident on two pedal phalanges (one proximal, one intermediate). Degenerative joint disease affects nearly all of the synovial joints, though not the temporomandibular joint. Both distal femora and well as the right navicular and talus are severely affected, with exposed trabecular bone.

Flo, as a well-known member of the Kasekela community, is a particularly apt case study supporting the hypothesis that older chimpanzees exhibit more skeletal pathologies, since they have had longer to accumulate them (Jurmain 1989, 1997). The next chapter tests this demographic trend, reports which body regions are most commonly affected by skeletal lesions, and explores other possible connections between skeletal lesions and dominance rank, sex, and cause of death.

References

Anestis SF (2005) Behavioral style, dominance rank, and urinary cortisol in young chimpanzees (*Pan troglodytes*). Behaviour 142:1251–1274

Anestis SF, Bribiescas RG, Hasselschwert DL (2006) Age, rank, and personality effects on the cortisol sedation stress response in young chimpanzees. Physiol Behav 89:287–294

Arcadi AC, Wrangham RW (1999) Infanticide in Chimpanzees: review of cases and a new within-group observation from the Kanyawara Study Group in Kibale National Park. Primates 40(2):337–351

Aufderheide AC, Rodriguez-Martín C (1998) The Cambridge Encyclopedia of Human Paleopathology. Cambridge University Press, Cambridge

Barfield WR, Holmes RE, Hartsock LA (2017) Heterotopic ossification in trauma. Orthop Clin 48(1):35–46

Bergman TJ, Beehner JC, Cheney DL, Seyfarth RM, Whitten PL (2005) Correlates of stress in free-ranging male chacma baboons, *Papio hamadryas* ursinus. Anim Behav 70:703–713

Buikstra JE, Ubelaker DH (1994) Standards for Data Collection from Human Skeletal Remains. Arkansas Archaeological Survey, Fayetteville

Bygott JD (1979) Agonisitc behavior, dominance and social structure in wild chimpanzees of the Gombe National Park. In: Hamburg DA, McCown ER (eds) The Great Apes. Benjamin/ Cummings Publishing Company, Menlo Park, pp 405–427

Carter M, Pontzer H, Wrangham RW, Peterhans K (2008) Skeletal pathology in *Pan troglodytes schweinfurthii* in Kibale National Park, Uganda. Am J Phys Anthropol 135(4):389–403

Chichinadze K, Chichinadze N (2008) Stress-induced increase of testosterone: contributions of social status and sympathetic reactivity. Physiol Behav 94(4):595–603

Chrcanovic BR, Abreu MHNG, Custódio ALN (2010) Prevalance of enamel pearls in teeth from a human teeth bank. J Oral Sci 52(2):257–260

Collins, CA (2016) Gombe Stream Research Center Annual Report. Personal communication

Constable JL, Ashley MV, Goodall J, Pusey AE (2001) Noninvasive paternity assignment in Gombe chimpanzees. Mol Ecol 10(5):1279–1300

Foster MW, Gilby IC, Murray CM, Johnson A, Wroblewski EE, Pusey AE (2009) Alpha male chimpanzee grooming patterns: implications for dominance "style". Am J Primatol 71(2):136–144

Galloway A, Morbeck ME, Zihlman AL (1996) Sex differences in the vertebral column of Gombe chimpanzees. Primates 37:443–453

Galloway A, Wedel VL (2014) Bones of the skull, the dentition, and osseous structures of the throat. In: Wedel VL, Galloway A (eds) Broken Bones: Anthropological Analysis of Blunt Force Trauma. Charles C. Thomas, Springfield, pp 133–160

Goodall J (1977) Infant killing and cannibalism in free-living chimpanzees. Folia Primatol 28:259–282

Goodall J (1986) The Chimpanzees of Gombe: Patterns of Behavior. Harvard University Press, Cambridge

Goodall J (1992) Unusual violence in the overthrow of an alpha male chimpanzee at Gombe. In: Nishida T, McGrew WC, Marler P, Pickford M, FBM dW (eds) Topics in Primatology, Human Origins, vol 1. University of Tokyo Press, Tokyo, pp 131–142

Harman MP (2005) Primate Collection in the Powell-Cotton Museum. Powell-Cotton Museum, Birchington

Indriati E, Ant6n SC (2010) The calvaria of Sangiran 38, Sendangbusik, Sangiran Dome, Java. HOMO 61(4):225–243

Jameson KA, Appleby MC, Freeman LC (1999) Finding an appropriate order for a hierarchy based on probabilistic dominance. Anim Behav 57:991–998

Jurmain R (1989) Trauma, degenerative disease, and other pathologies among the Gombe chimpanzees. Am J Phys Anthropol 80:229–237

Jurmain R (1997) Skeletal evidence of trauma in African apes, with special reference to the Gombe chimpanzees. Primates 38(1):1–14

Jurmain R (2000) Degenerative joint disease in African great apes: an evolutionary perspective. J Hum Evol 39:185–203

Jurmain R, Kilgore L (1998) Sex-related patterns of trauma in humans and great apes. In: Grauer AL, Stuart-MacAdam P (eds) Sex and Gender in Paleopathological Perspective. Cambridge University Press, Cambridge, pp 11–26

Jurmain RD, Kilgore L (1995) Skeletal evidence of osteoarthritis - a paleopathological perspective. Ann Rheum Dis 54(6):443–450

Kahlenberg SM, Thompson ME, Muller MN, Wrangham RW (2008) Immigration costs for female chimpanzees and male protection as an immigrant counterstrategy to intrasexual aggression. Anim Behav 76:1497–1509

Kawanaka K (1990) Alpha males' interactions and social skills. In: Nishida T (ed) The chimpanzees of the Mahale Mountains: sexual and life history strategies. University of Tokyo Press, Tokyo, pp 171–187

Keele BF, Holland Jones J, Terio KA, Estes JD, Rudicell RS, Wilson ML, Li Y, Learn GH, Beasley TM, Schumacher-Stankey J, Wroblewski E, Mosser A, Raphael J, Kamenya S, Lonsdorf EV, Travis DA, Mlengeya T, Kinsel MJ, Else JG, Silvestri G, Goodall J, Sharp PM, Shaw GM, Pusey AE, Hahn BH (2009) Increased mortality and AIDS-like immunopathology in wild chimpanzees infected with SIVcpz. Nature 460:515–519

Key CA, Aiello LC, Molleson T (1994) Cranial suture closure and its implications for age estimation. Int J Osteoarchaeol 4:193–207

Kilgore L (1989) Dental pathologies in ten free-ranging chimpanzees from Gombe National Park, Tanzania. Am J Phys Anthropol 80:219–227

Kirchhoff CA (2010) From Birth to Bones: Skeletal Evidence for Health, Disease, and Injury in the Gombe Chimpanzees. Ph.D. Thesis, University of Minnesota, Minneapolis

Kirchhoff CA, Wilson ML, Mjungu DC, Raphael J, Kamenya S, Collins DA (2018) Infanticide in chimpanzees: Taphonomic case studies from Gombe. Am J Phys Anthropol 165(1):108–122

Lane J (2006) Can non-invasive glucocorticoid measures be used as reliable indicators of stress in animals. Anim Welf 15:331–342

Lovell NC (1990) Patterns of injury and illness in great apes: a skeletal analysis. Smithsonian Institution Press, Washington, DC

Lovell NC (1991) An evolutionary framework for assessing illness and injury in nonhuman primates. Yearb Phys Anthropol 34:117–155

Miles A (1994) Non-union of the epiphysis of the acromion in the skeletal remains of a Scottish population of ca. 1700. Int J Osteoarchaeol 4(2):149–163
Morbeck ME (1987) Body size and proportion in Gombe chimpanzees. Am J Phys Anthropol 72:234
Morbeck ME, Galloway A, Mowbray KM, Zihlman AL (1994) Skeletal asymmetry and hand preference during termite fishing by Gombe chimpanzees. Primates 35(1):99–103
Morbeck ME, Galloway A, Sumner DR (2002) Getting old at Gombe: skeletal aging in wild-ranging chimpanzees. In: Erwin JM, Hof PR (eds) Aging in Non-human Primates. Karger, Basel, pp 48–62
Morbeck ME, Galloway A, Zihlman AL (1992) Gombe chimpanzee sex differences in the pelvis and observations of pubic and preauricular areas. Primates 33(1):129
Morbeck ME, Zihlman AL (1989) Body size and proportions in chimpanzees, with special reference to *Pan troglodytes schweinfurthii* from Gombe National Park, Tanzania. Primates 30(3):369–382
Morbeck ME, Zihlman AL, Sumner DR, Galloway A (1991) Poliomyelitis and skeletal asymmetry in Gombe chimpanzees. Primates 32(1):77–91
Muller MN, Wrangham RW (2004) Dominance, cortisol and stress in wild chimpanzees (*Pan troglodytes schweinfurthii*). Behav Ecol Sociobiol 55:332–340
Murray CM (2007) Method for assigning categorical rank in female *Pan troglodytes schweinfurthii* via the frequency of approaches. Int J Primatol 28(4):853–864
Murray CM, Eberly LE, Pusey AE (2006) Foraging strategies as a function of season and rank among wild female chimpanzees (*Pan troglodytes*). Behav Ecol 17(6):1020–1028
Neidich DL (2014) A Comparative Study of Upper Limb Mechânical Stress in the Pre-Colombian Tennessee River Valley. M.S. Thesis, Illinois State University, Normal
Novak SA, Hatch MA (2009) Intimate wounds: craniofacial trauma in women and female chimpanzees. In: Muller MN, Wrangham RW (eds) Sexual Coercion in Primates and Humans: An Evolutionary Perspective on Male Aggression Against Females. Harvard University Press, Cambridge, pp 322–345
Ortner DJ (2003) Identification of Pathological Conditions in Human Skeletal Remains, 2nd edn. Academic, Amsterdam
Pobiner BL, DeSilva J, Sanders WJ, Mitani JC (2007) Taphonomic analysis of skeletal remains from chimpanzee hunts at Ngogo, Kibale National Park, Uganda. J Hum Evol 52(6):614–636
Pusey AE, Oehlert GW, Williams JM, Goodall J (2005) Influence of ecological and social factors on body mass of wild chimpanzees. Int J Primatol 26(1):3–31
Pusey AE, Williams J, Goodall J (1997) The influence of dominance rank on reproductive success of female chimpanzees. Science 277:828–831
Rando C, Waldron T (2012) TMJ osteoarthritis: A new approach to diagnosis. Am J Phys Anthropol 148:45–53
Ray JC, Sapolsky RM (1992) Styles of male social behavior and their endocrine correlates among high-ranking wild baboons. Am J Primatol 28:231–250
Resnick D (1988) Diagnosis of Bone and Joint Disorders. Saunders, Philadelphia
Robbins MM, Gray M, Fawcett KA, Nutter FB, Uwingeli P, Mburanumwe I, Kagoda E, Basabose A, Stoinski TS, Cranfield MR (2011) Extreme conservation leads to recovery of the Virunga mountain gorillas. PLoS One 6(6):e19788
Roberts AM, Peters TJ, Brown KR (2007) New light on old shoulders: Palaeopathological patterns of arthropathy and enthesopathy in the shoulder complex. J Anat 211(4):485–492
Rogers J, Waldron T (1995) A Field Guide to Joint Disease in Archaeology. Wiley, Chichester
Santiago ML, Rodenburg CM, Kamenya S, Bibollet-Ruche F, Gao F, Bailes E, Meleth S, Soong SJ, Kilby JM, Moldoveanu Z, Fahey B, Muller MN, Ayouba A, Nerrienet E, McClure HM, Heeney JL, Pusey AE, Collins DA, Boesch C, Wrangham RW, Goodall J, Sharp PM, Shaw GM, Hahn BH (2002) SIVcpz in wild chimpanzees. Science 295(5554):465–465
Sapolsky RM (1983) Endocrine aspects of social instability in the Olive baboon (*Papio anubis*). Am J Primatol 5:365–379

Sapolsky RM (2005) The influence of social hierarchy on primate health. Science 308(5722): 648–652

Scheuer L, Black S (2004) The Juvenile Skeleton. Academic, Amsterdam

Schino G, Rosati L, Geminiani S, Aureli F (2007) Post-conflict anxiety in Japanese macaques (*Macaca fuscata*): aggressor's and victim's perspectives. Ethology 113(11):1081–1088

Stirland AJ (2001) Raising the Dead: The Skeleton Crew of King Henry VIII's Great Ship, the Mary Rose. Wiley, Chichester

Sumner DR, Morbeck ME, Lobick JJ (1989) Apparent age-related bone loss among adult female Gombe chimpanzees. Am J Phys Anthropol 79:225–234

Takahata Y (1990) Social relationships among adult males. In: Nishida T (ed) The chimpanzees of the Mahale Mountains: sexual and life history strategies. Tokyo University Press, Tokyo, pp 149–170

Terio KA, Kinsel MJ, Raphael J, Mlengeya T, Lipende I, Kirchhoff CA, Gilagiza B, Wilson ML, Kamenya S, Estes JD, Keele BF, Rudicell RS, Liu W, Patton S, Collins A, Hahn BH, Travis DA, Lonsdorf EV (2011) Pathologic lesions in chimpanzees (*Pan trogylodytes schweinfurthii*) from Gombe National Park, Tanzania, 2004–2010. J Zoo Wildl Med 42(4):597–607

Thompson K (2007) Diseases of bones. In: Maxie M (ed) Jabb, Keeny and Palmer's Pathology of Domestic Animals, vol 1. Saunders, Philadelphia

Thompson ME, Muller MN, Wrangham RW, Lwanga JS, Potts KB (2009) Urinary C-peptide tracks seasonal and individual variation in energy balance in wild chimpanzees. Horm Behav 55:299–305

Virgin J, Charles E, Sapolsky RM (1997) Styles of male social behavior and their endocrine correlates among low-ranking baboons. Am J Primatol 42:25–39

Weisbrode SE (2007) Bones and joints. In: McGavin MD (ed) Pathologic Basis of Veterinary Disease. Mosby, St. Louis

Williams JM, Lonsdorf EV, Wilson ML, Schumacher-Stankey J, Goodall J, Pusey AE (2008) Causes of death in the Kasekela chimpanzees of Gombe National Park, Tanzania. Am J Primatol 70:766–777

Wilson ML, Wallauer WR, Pusey AE (2004) New cases of intergroup violence among chimpanzees in Gombe National Park, Tanzania. Int J Primatol 25(3):523–549

Wrangham RW, Wilson ML, Muller MN (2006) Comparative rates of violence in chimpanzees and humans. Primates 47:14–26

Wroblewski EE, Murray CM, Keele BF, Schumacher-Stankey JC, Hahn BH, Pusey AE (2009) Male dominance rank and reproductive success in chimpanzees, *Pan troglodytes schweinfurthii*. Anim Behav 77:873–885

Yochum TR, Rowe LT (2005) Arthritic disorders. In: Yochum TR, Rowe LT (eds) Yochum and Rowe's Essentials of Skeletal Radiology, 3rd edn. Lippincot, Williams, and Wilkins, Philadelphia

Zihlman AL, Bolter D, Boesch C (2004a) Skeletal and dental development in wild chimpanzees from Taï National Forest, Ivory Coast and Gombe Reserve, Tanzania. Am J Phys Anthropol 123(S38):215–216

Zihlman AL, Bolter D, Boesch C (2004b) Wild chimpanzee dentition and its implications for assessing life history in immature hominin fossils. Proc Natl Acad Sci 101(29):10541–10543

Zihlman AL, Morbeck ME, Goodall J (1990) Skeletal biology and individual life history of Gombe chimpanzees. J Zool Soc Lond 221:37–61

Chapter 2
Analysis of Skeletal Lesions

2.1 Introduction

Because dominance rank plays an important role in the social lives of chimpanzees, I examined possible relationships between rank, age, sex, and skeletal lesions in the Gombe chimpanzee skeletal collection. From behavioral studies, we know that dominance rank affects access to resources (Pusey et al. 2005; Murray et al. 2006), inter-individual relationships (Foster et al. 2009; Goodall 1986) and reproductive success (Pusey et al. 1997; Constable et al. 2001; Wroblewski et al. 2009), among other variables. As such, dominance rank can be considered a useful proxy for measuring individual fitness. How chimpanzee fitness may be affected by trauma and pathology is relevant for better understanding the selective pressures confronting chimpanzees.

The long-term behavioral research at Gombe, as well as studies of other wild primate skeletons, allows for reasonable predictions concerning the relationship between rank and trauma and pathology. Males who achieved high rank, because of the agonistic encounters necessary for securing and maintaining high rank (Muller and Wrangham 2004; Thompson et al. 2009), should have the highest frequencies of trauma. Low-ranking females, particularly peripheral or immigrant females, should also have a high incidence of trauma because of elevated levels of intrasexual aggression (Kahlenberg et al. 2008). Previous studies of wild primate skeletal material (Jurmain 1997; Jurmain and Kilgore 1998; Latimer 1993; Lovell 1990) have found a correlation between prevalence of trauma and pathology and age, and I expect those trends to hold true for this enlarged sample from Gombe, as a longer life span means a chimpanzee has more time to accumulate injuries and illnesses. Degenerative pathologies such as antemortem tooth loss and arthropathy should be closely correlated with age, and are likely to be less influenced by dominance rank. Most studies of primate skeletons focus on adults, sometimes including older subadults. As this sample includes a much wider variety of age categories (from young

C. A. Kirchhoff, *Life and Death in the Gombe Chimpanzees*,
Developments in Primatology: Progress and Prospects,
https://doi.org/10.1007/978-3-030-18355-4_2

infant to old adult), the correlation between age and trauma incidence may be less strong compared to previous findings. This is especially relevant in light of the injuries observed during chimpanzee infanticides (Arcadi and Wrangham 1999; Kirchhoff et al. 2018) and the inclusion of four infanticide victims in this sample.

2.2 Materials and Methods

There are 49 chimpanzees included in the skeletal sample from Gombe presented in this book (see Table 1.1). Of these, 36 are complete (or nearly) skeletons suitable for quantitative analysis of skeletal lesions (16 females, 19 males, 1 unknown sex). In addition, two males with incomplete post-crania (Cusano, Tumaini) are included in the analyses of craniodental lesions, and Gremlin's male infant (lacking ribs) is included in the regional analyses of lesion frequency, excluding the thorax. Table 2.1a, 2.1b, and 2.1c summarize the skeletal lesion and demographic data of these 36 chimpanzees. Observational criteria for skeletal lesions are outlined in Chap. 1. This sample provides the basis for a quantitative analysis of skeletal trauma and pathology, examining the patterning of lesions, testing for age-mediated variation, possible differences between sex and dominance rank categories, and presenting comparisons with other work.

Because many chimpanzees are missing a few smaller bones (usually from the hands/feet), trauma and pathology rates are expressed as a percentage of observable bones affected by trauma or pathology, in order to control for the differences in number of preserved skeletal elements. I also examined frequency of lesions by body region, with regions defined as follows: (1) head (skull and teeth), (2) trunk (vertebrae, ribs, sternum, hyoid, scapulae, clavicles, pelvis), (3) forelimb (humerus, radius, ulna), (4) hindlimb (femur, patella, tibia, fibula), and (5) hand/foot or distal limb (podials, metapodials, phalanges).

I include summary statistics on which elements are most often affected by lesions. I also employed statistical tests, carried out using SPSS version 24.0 (IBMCorp 2016) or ARC (Cook and Weisberg 1999) for the regression analyses, and the details of each analysis are presented below. I carried out Mann–Whitney (also known as rank-sum or Wilcoxon) tests and z-score tests (for the difference between population proportions (Devore and Peck 2001)) to examine sex differences in skeletal lesions. I employed regression analysis to examine possible relationships between skeletal lesion rates and age at death, and evaluated possible differences in skeletal lesion rates between dominance rank categories using Kruskal–Wallis tests. I also tested for a relationship between age and dominance rank using a Kruskal–Wallis test. I calculated effects sizes for all non-parametric tests after Rosenthal and DiMatteo (2001).

Table 2.1a Skeletal lesion data for 36 Gombe chimpanzees with analyzable skeletons

Chimpanzee	Age (years)	Sex	Observed bones	Number of traumatic lesions	Number of perimortem traumatic lesions	Number of antemortem traumatic lesions	Number pathologic lesions	Number of any lesion	Cause of death	Trauma rate (number of lesions/observed bones)	Pathology rate (number of lesions/observed bones)	Rate of any lesion (number of lesions/observed bones)	Mother's rank at time of individual's birth (Mother's rank)	Highest achieved rank (Highest rank)	Rank at time of death (Death rank)
Andromeda	0.73	F	132	54	54	0	0	54	Traumatic	0.41	0.00	0.41	2	3	3
Atlas	31.30	M	216	5	0	5	27	32	Non-traumatic	0.02	0.13	0.15	2	1	3
Bwavi male	40	M	229	9	0	9	53	62	Non-traumatic	0.04	0.23	0.27	n/a	n/a	n/a
Charlie	25.87	M	183	3	2	1	12	15	Traumatic	0.02	0.07	0.08	n/a	1	1
Ebony	8.19	M	206	3	0	3	7	10	Traumatic	0.01	0.03	0.05	n/a	3	3
Echo	34.00	F	210	12	0	12	18	30	Traumatic	0.06	0.09	0.14	n/a	n/a	n/a
Flint	8.50	M	181	0	0	0	7	7	Non-traumatic	0.00	0.04	0.04	1	3	3
Flo	53.14	F	188	9	0	9	31	40	Non-traumatic	0.05	0.16	0.21	n/a	1	1
Fred	0.90	M	141	0	0	0	14	14	Non-traumatic	0.00	0.10	0.10	1	3	3
Gaia's infant	0.00	M	125	0	0	0	0	0	Non-traumatic	0.00	0.00	0.00	2	3	3
Galahad	11.84	M	170	0	0	0	14	14	Non-traumatic	0.00	0.08	0.08	2	3	3
Gilka	18.87	F	177	2	0	0	23	25	Non-traumatic	0.01	0.13	0.14	2	3	3
Goblin	39.96	M	208	17	0	17	38	55	Non-traumatic	0.08	0.18	0.26	2	1	1
Groucho	1.30	M	121	0	0	0	1	1	Non-traumatic	0.00	0.01	0.01	2	3	3

(continued)

Table 2.1a (continued)

Chimpanzee	Age (years)	Sex	Observed bones	Number of traumatic lesions	Number of perimortem traumatic lesions	Number of antemortem traumatic lesions	Number pathologic lesions	Number of any lesion	Cause of death	Trauma rate (number of lesions/ observed bones)	Pathology rate (number of lesions/ observed bones)	Rate of any lesion (number of lesions/ observed bones)	Mother's rank at time of individual's birth (Mother's rank)	Highest achieved rank (Highest rank)	Rank at time of death (Death rank)
Gyre	0.80	M	111	2	0	2	0	2	Non-traumatic	0.02	0.00	0.02	2	3	3
Hugo	38.58	M	183	8	0	8	33	41	Non-traumatic	0.04	0.18	0.22	n/a	n/a	3
Jackson	10.41	M	199	0	0	0	10	10	Non-traumatic	0.00	0.05	0.05	3	3	3
Jomeo	30.87	M	198	4	0	4	26	30	Non-traumatic	0.02	0.13	0.15	n/a	1	1
Kidevu	25.59	F	205	0	0	0	9	9	Non-traumatic	0.00	0.04	0.04	n/a	1	2
MacDee	13.50	M	196	5	4	1	18	23	Traumatic	0.03	0.09	0.12	n/a	3	3
Madam Bee	28.22	F	197	12	8	4	8	20	Traumatic	0.06	0.04	0.10	n/a	n/a	n/a
Mel	10.71	M	203	3	0	3	14	17	Traumatic	0.01	0.07	0.08	1	3	3
Melissa	36.31	F	207	3	0	3	15	18	Non-traumatic	0.01	0.07	0.09	n/a	2	2
Melissa's infant	0.00	M	99	0	0	0	0	0	Non-traumatic	0.00	0.00	0.00	1	3	3
Michaelmas	12.99	M	201	4	0	4	4	8	Non-traumatic	0.02	0.02	0.04	2	3	3
Miff	30.90	F	202	1	0	1	17	18	Non-traumatic	0.00	0.08	0.09	2	1	1
Old Female	50	F	198	6	0	6	18	24	Traumatic	0.03	0.09	0.12	n/a	n/a	n/a
Pallas	27.20	F	165	0	0	0	10	10	Non-traumatic	0.00	0.06	0.06	n/a	3	3

Passion	30.61	F	225	9	0	9	25	34	Non-traumatic	0.04	0.11	0.15	n/a	1	1
Patti	44.25	F	214	5	1	4	11	16	Traumatic	0.02	0.05	0.07	n/a	1	1
Patti's infant	0.00	n/a	125	0	0	0	0	0	Non-traumatic	0.00	0.00	0.00	3	3	3
Rejea	0.27	F	106	40	40	0	0	40	Traumatic	0.38	0.00	0.38	n/a	3	3
Sherehe	15.78	F	215	1	1	0	13	14	Non-traumatic	0.00	0.06	0.07	3	2	2
Sugar	10.91	F	178	0	0	0	6	6	Non-traumatic	0.00	0.03	0.03	n/a	3	3
Vincent	28.47	M	195	60	26	34	31	91	Traumatic	0.31	0.16	0.47	n/a	1	3
Yolanda	24.00	F	216	3	0	3	26	29	Non-traumatic	0.01	0.12	0.13	n/a	2	2

Table 2.1b Skeletal lesion rate data for 36 Gombe chimpanzees with analyzable skeletons

Chimpanzee	Age (years)	Sex	Trauma rate (number of lesions/observed bones)	Pathology rate (number of lesions/observed bones)	Rate of any lesion (number of lesions/observed bones)
Andromeda	0.73	F	0.41	0.00	0.41
Atlas	31.30	M	0.02	0.13	0.15
Bwavi male	40	M	0.04	0.23	0.27
Charlie	25.87	M	0.02	0.07	0.08
Ebony	8.19	M	0.01	0.03	0.05
Echo	34.00	F	0.06	0.09	0.14
Flint	8.50	M	0.00	0.04	0.04
Flo	53.14	F	0.05	0.16	0.21
Fred	0.90	M	0.00	0.10	0.10
Gaia's infant	0.00	M	0.00	0.00	0.00
Galahad	11.84	M	0.00	0.08	0.08
Gilka	18.87	F	0.01	0.13	0.14
Goblin	39.96	M	0.08	0.18	0.26
Groucho	1.30	M	0.00	0.01	0.01
Gyre	0.80	M	0.02	0.00	0.02
Hugo	38.58	M	0.04	0.18	0.22
Jackson	10.41	M	0.00	0.05	0.05
Jomeo	30.87	M	0.02	0.13	0.15
Kidevu	25.59	F	0.00	0.04	0.04
MacDee	13.50	M	0.03	0.09	0.12
Madam Bee	28.22	F	0.06	0.04	0.10
Mel	10.71	M	0.01	0.07	0.08
Melissa	36.31	F	0.01	0.07	0.09
Melissa's infant	0.00	M	0.00	0.00	0.00
Michaelmas	12.99	M	0.02	0.02	0.04
Miff	30.90	F	0.00	0.08	0.09
Old Female	50	F	0.03	0.09	0.12
Pallas	27.20	F	0.00	0.06	0.06
Passion	30.61	F	0.04	0.11	0.15
Patti	44.25	F	0.02	0.05	0.07
Patti's infant	0.00	n/a	0.00	0.00	0.00
Rejea	0.27	F	0.38	0.00	0.38
Sherehe	15.78	F	0.00	0.06	0.07
Sugar	10.91	F	0.00	0.03	0.03
Vincent	28.47	M	0.31	0.16	0.47
Yolanda	24.00	F	0.01	0.12	0.13

Table 2.1c Demographic data for 36 Gombe chimpanzees with analyzable skeletons

Chimpanzee	Age (years)	Sex	Cause of death	Mother's rank at time of individual's birth (Mother's rank)	Highest achieved rank (Highest rank)	Rank at time of death (Death rank)
Andromeda	0.73	F	Traumatic	2	3	3
Atlas	31.30	M	Non-traumatic	2	1	3
Bwavi male	40	M	Non-traumatic	n/a	n/a	n/a
Charlie	25.87	M	Traumatic	n/a	1	1
Ebony	8.19	M	Traumatic	n/a	3	3
Echo	34.00	F	Traumatic	n/a	n/a	n/a
Flint	8.50	M	Non-traumatic	1	3	3
Flo	53.14	F	Non-traumatic	n/a	1	1
Fred	0.90	M	Non-traumatic	1	3	3
Gaia's infant	0.00	M	Non-traumatic	2	3	3
Galahad	11.84	M	Non-traumatic	2	3	3
Gilka	18.87	F	Non-traumatic	2	3	3
Goblin	39.96	M	Non-traumatic	2	1	1
Groucho	1.30	M	Non-traumatic	2	3	3
Gyre	0.80	M	Non-traumatic	2	3	3
Hugo	38.58	M	Non-traumatic	n/a	n/a	3
Jackson	10.41	M	Non-traumatic	3	3	3
Jomeo	30.87	M	Non-traumatic	n/a	1	1
Kidevu	25.59	F	Non-traumatic	n/a	1	2
MacDee	13.50	M	Traumatic	n/a	3	3
Madam Bee	28.22	F	Traumatic	n/a	n/a	n/a
Mel	10.71	M	Traumatic	1	3	3
Melissa	36.31	F	Non-traumatic	n/a	2	2
Melissa's infant	0.00	M	Non-traumatic	1	3	3

(continued)

Table 2.1c (continued)

Chimpanzee	Age (years)	Sex	Cause of death	Mother's rank at time of individual's birth (Mother's rank)	Highest achieved rank (Highest rank)	Rank at time of death (Death rank)
Michaelmas	12.99	M	Non-traumatic	2	3	3
Miff	30.90	F	Non-traumatic	2	1	1
Old Female	50	F	Traumatic	n/a	n/a	n/a
Pallas	27.20	F	Non-traumatic	n/a	3	3
Passion	30.61	F	Non-traumatic	n/a	1	1
Patti	44.25	F	Traumatic	n/a	1	1
Patti's infant	0.00	n/a	Non-traumatic	3	3	3
Rejea	0.27	F	Traumatic	n/a	3	3
Sherehe	15.78	F	Non-traumatic	3	2	2
Sugar	10.91	F	Non-traumatic	n/a	3	3
Vincent	28.47	M	Traumatic	n/a	1	3
Yolanda	24.00	F	Non-traumatic	n/a	2	2

2.3 Is the Skeletal Sample Representative of the Overall Population at Gombe?

To assess the representativeness of the skeletal sample, I made two comparisons. First, to the demographics of the current population at Gombe, according to a recent report from the Gombe Stream Research Center (D. Anthony Collins, personal communication, 2016). I also contrasted the skeletal sample with the record of all deaths in the Kasekela community published by Williams et al. (2008).

In 2016, there were 77 living chimpanzees observed in the Kasekela and Mitumba communities (GSRC). Tables 2.2 and 2.3 present contrasting data for the demographics of the skeletal sample versus the current, living population of chimpanzees from Gombe. There are several important differences worth noting between the skeletal sample versus living population, including that there are no juveniles included in the skeletal sample, while this demographic category made up 12% of the chimpanzee population at Gombe in 2016. Similarly, adults make up a smaller proportion of the skeletal sample (37%) compared to the living population (48%). This may reflect, in part, the high mortality rates of young chimpanzees (Goodall 1986).

Table 2.2 Demographic summary of the chimpanzee skeletal sample from Gombe

Age category (Goodall 1986)	Females	Males	Unknown sex	Total
Infant (0–4 years)	3 (6%)	9 (18%)	1 (2%)	12 (27%)
Juvenile (5–7 years)	0	0	0	0
Adolescent (females: 8–14 or 15, males: 8–15)	1 (2%)	7 (14%)	0	8 (16%)
Adult (females: 14 or 15–33, males: 16–33)	9 (18%)	9 (18%)	0	18 (37%)
Old adult (34+)	5 (10%)	5 (10%)	0	10 (20%)
Total	14 (37%)	30 (61%)	1 (2%)	49

Percentages in parentheses indicate the proportion of the total sample ($n = 49$) made up by each demographic category

Table 2.3 Demographic summary of living chimpanzees at Gombe as of late 2016 (GSRC)

Age category (Goodall 1986)	Females	Males	Total
Infant (0–4 years)	7 (9%)	9 (11%)	16 (21%)
Juvenile (5–7 years)	4 (5%)	5 (6%)	9 (12%)
Adolescent (females: 8–14 or 15, males: 8–15)	7 (9%)	3 (4%)	10 (13%)
Adult (females: 14 or 15–33, males: 16–33)	25 (32%)	12 (16%)	37 (48%)
Old adult (34+)	5 (6%)	0	5 (6%)
Total	48 (62%)	29 (38%)	77

Data for the Mitumba and Kasekela communities are combined (as are the data for the skeletal collection). Percentages in parentheses indicate the proportion of the total sample ($n = 77$) made up by each demographic category

Of greatest interest are the nearly inverse sex ratios of the skeletal sample versus living population. While males comprise 61% of the skeletal sample, they make up 38% of the living population. In contrast, females make up 37% of the skeletal sample, while 62% of the living population is female. This may be related to higher mortality rates among male chimpanzees compared to females (Hill et al. 2001).

While the skeletal sample does not closely represent the population of chimpanzees currently living in Gombe National Park, the skeletal collection may be a reasonable representation of chimpanzees who have died. Table 2.4 presents a demographic summary of chimpanzees from Kasekela who are known to have died since the start of human observation at Gombe in 1960 through 2006 (Williams et al. 2008). Distribution across age categories in the data set examined by Williams et al. (2008) and the skeletal sample is similar (see Table 2.2 for skeletal sample), but the sex ratios again differ. While the proportion of males and females who died is approximately equal (Table 2.4), females comprise only 37% of the skeletal sample. This may be related to differences in foraging patterns between male and female chimpanzees. With females more likely to disperse or forage alone (or only with dependent offspring) (Goodall 1986; Wrangham 2000), their bodies may be less likely to be recovered at the time of death compared to males.

Comparing frequencies of causes of death between samples provides another useful means of assessing the degree to which the skeletal sample can be considered

Table 2.4 Demographic summary of chimpanzees from Kasekela who died between 1960 and 2006

Age category	Females	Males	Unknown sex	Total
Infant (0–4 years)	20 (12%)	29 (18%)	8 (5%)	57 (35%)
Juvenile (5–7 years)	4 (2%)	3 (2%)		7 (4%)
Adolescent (females: 8–14 or 15, males: 8–15)	15 (9%)	11 (7%)		26 (16%)
Adult (females: 14 or 15–33, males: 16–33)	31 (19%)	21 (13%)		52 (32%)
Old adult (34+)	10 (6%)	9 (6%)		19 (12%)
Total	80 (50%)	73 (45%)		161

Percentages given are the percent of the total sample (n = 161) (Williams et al. 2008)

representative of chimpanzees from Gombe who have died. Table 2.5 gives a demographic summary according to cause of death (as defined by Williams et al. 2008), while Table 2.6 compares frequency of causes of death between the skeletal sample (n = 48 deceased chimpanzees with known sex) and the data set analyzed by Williams and colleagues. In Table 2.6, data from the skeletal sample are calculated according to the method described by Williams and colleagues: the percentage of deaths of unknown cause is calculated from the total number of deaths (n = 48), while the other frequencies are calculated as the percentage of deaths with known cause (e.g., of the 25 deaths ascribed to a known cause for male chimpanzees in the skeletal sample, 15 are attributable to illness = 60%).

Notable differences in the frequencies of causes of death between the skeletal sample and the Williams data set include that none of the female infants in the skeletal sample died as a result of orphaning, compared to 15% in the data set analyzed by Williams and colleagues. The frequency of female deaths without an attributable cause is also higher in the Williams data set (40%) compared to the skeletal sample (11%). This likely reflects both female foraging patterns, which make it more likely females will forage alone (Goodall 1986; Wrangham 2000) as well as the fact that a chimpanzee whose body has been recovered and skeleton has been preserved is more likely to have a known cause of death. Similarly, the frequency with which cause of death can be attributed to intraspecific aggression is higher in the skeletal sample (38%) compared to the Williams data set (15%). This is likely related (inversely) to the difference in the number of deaths of unknown cause: observed deaths due to conspecific aggression are more likely to involve recovery and preservation of the skeleton compared to a female chimpanzee who forages alone and becomes ill or disappears.

While the skeletal sample is not a perfect representation of the population of chimpanzees from Gombe who have died, it is not unreasonable to attempt to detect patterns in the skeletal sample that might tell us something about chimpanzees in general, or at Gombe in particular. Still, the skeletal sample represents only 49 total animals, and it was not possible to include all of these individuals in all of the quantitative analyses discussed below. This highlights the importance of long-term research on long-lived, socially complex animals such as chimpanzees: it is only

Table 2.5 Frequency of causes of death by demographic category in the skeletal sample

Skeletal sample			Age									
			0–5 years		5–10 years		10–20 years		20–30 years		Over 30 years	
Cause of death	Male total	Female total	Males	Females	Males	Females	Males	Females	Males	Females	Males	Females
Illness	15	8	4	0	0	0	4	2	1	2	6	4
Intraspecific aggression	5	6	1	3	1	0	1	0	2	1	0	2
Injury	1	2	0	0	0	0	0	1	1	0	0	1
Orphaning	1	0	0	0	1	0	0	0	0	0	0	0
Maternal disability	1	0	1	0	0	0	0	0	0	0	0	0
Poaching	2	0	1	0	0	0	0	0	1	0	0	0
Unknown	5	2	2	0	0	0	0	0	0	2	3	0
Total	30	18	9	3	2	0	5	3	5	5	9	7

Table 2.6 Contrasting the frequencies of causes of death in the skeletal sample versus the Williams et al. (2008) sample, all ages

Cause of death	Percent male deaths skeletal sample	Percent male deaths overall sample (Williams et al. 2008)	Percent female deaths skeletal sample	Percent female deaths overall sample (Williams et al. 2008)
Illness	60%	59%	50%	61%
Intraspecific aggression	20%	24%	**38%**	**15%**
Injury	4%	8%	13%	6%
Orphaning	4%	2%	**0%**	**15%**
Maternal disability	4%	6%	0%	3%
Poaching	8%	2%	0%	0%
Unknown	17%	25%	**11%**	**40%**

Bold indicates divergence between the skeletal sample and death sample

after nearly six decades of research that some of the patterns of behavior and interaction become apparent, particularly with regard to how they might be reflected in the hard tissues. The next sections provide quantitative data on lesions to the skeleton and dentition, and an examination of how lesion location and frequency might intersect with demographic categories and dominance rank.

2.4 What Percent of the Sample Exhibits Trauma and Pathology?

Of the 36 chimpanzees included in the quantitative analysis, 25 (70%) have at least one skeletal lesion attributable to trauma, and 30 (83%) have at least one pathological (non-arthropathy) skeletal lesion. Trauma rates range from 0% of the skeletal elements affected to 41%, and pathology rates range from 0 to 23% (Table 2.1a, 2.1b, and 2.1c).

2.5 What Are the Most Frequently Affected Elements?

Of the 563 skeletal lesions attributable to pathology, the region most commonly affected encompasses the distal limbs (hand/foot = 33%, see Fig. 2.1). Scaling to determine how many of the observable elements were affected by pathologic lesions, the head was the region most commonly affected, with 78% of the observable elements affected (Fig. 2.2).

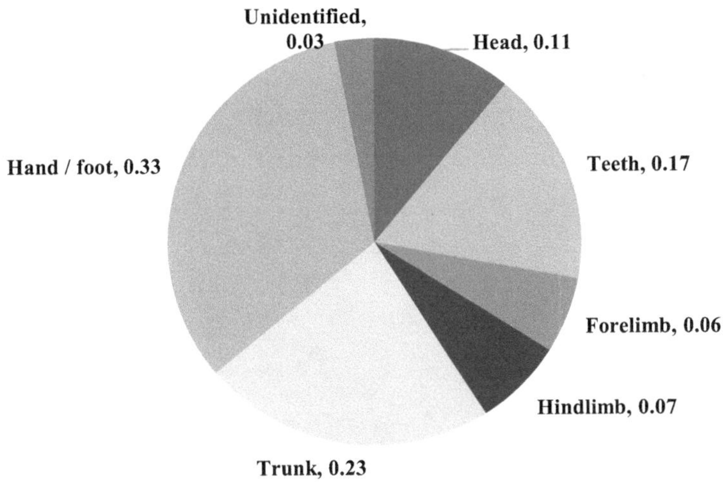

Fig. 2.1 Frequency of pathologic lesions by region

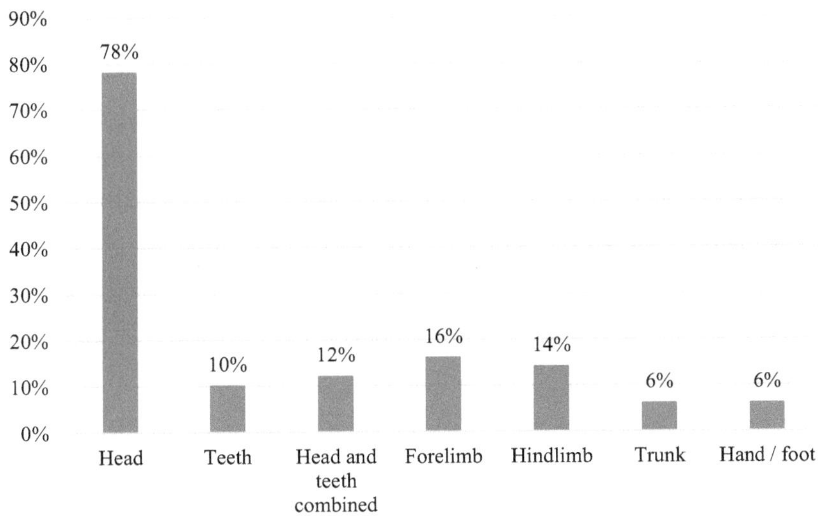

Fig. 2.2 Percentage of observed elements affected by pathologic lesions by region

Of the 315 traumatic lesions observed in this collection, the lesions to the trunk make up the greatest proportion (36%), while the distal extremities make up nearly another third (31%) (Fig. 2.3). Scaling to account for number of observable elements per region, the head was the most frequently affected by traumatic lesions, with 36% of these elements exhibiting skeletal evidence for trauma (Fig. 2.4).

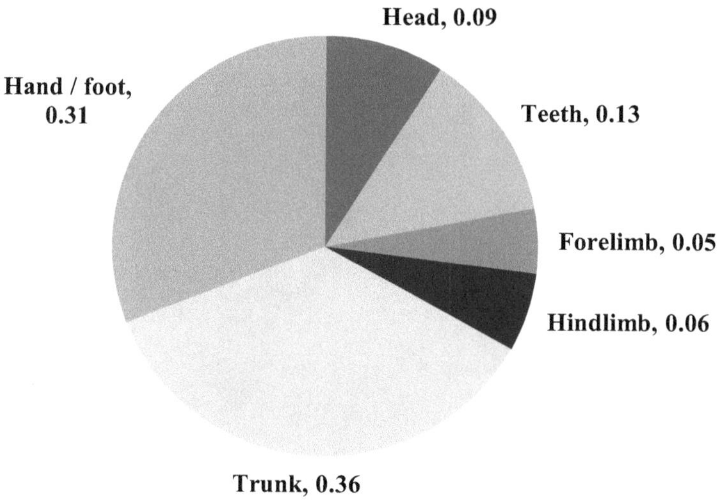

Fig. 2.3 Frequency of traumatic lesions by region

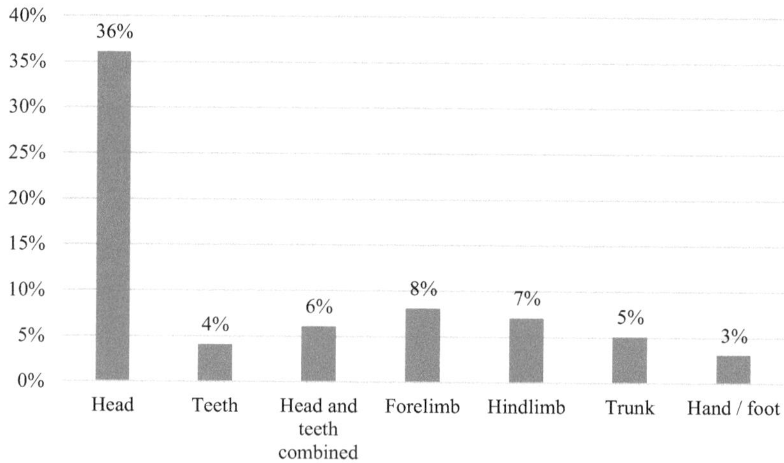

Fig. 2.4 Percentage of skeletal elements affected by trauma, by region

2.6 Arthropathy

Joint disease is considered separately from other types of skeletal lesions. Data presented in this segment are not also presented elsewhere. That is, arthropathy can result in both bone formation and/or bone destruction, but lesions counted here are not also considered "bone loss" or "bone formation" pathologies (Table 1.4).

Adults with complete skeletons and observable joint surfaces were included in this analysis ($n = 20$, 10 females, 10 males). Rix is an articulated skeleton housed at the University of Dar es Salaam and his joint surfaces could not be observed. Cusano's skeleton exhibits diagenic change from being buried for a long period of time as well as the postmortem loss of most of the right lower limb. His joints are discussed qualitatively in his narrative in Chap. 1. Flo's skeleton has desiccated cartilage adhering to her joint surfaces, which did not make them fully observable and comparable to the other animals in the sample. Her case is also discussed qualitatively in her narrative.

Due to the small sample, criteria for arthropathy/degenerative joint disease from Buikstra and Ubelaker (1994) were condensed into presence vs. absence for each joint. After Rando and Waldron (2012), the presence of lipping or osteophytes and increased porosity of articular bone was counted as arthropathy being present. If only one of those signs was present, arthropathy was scored as absent (in no cases were osteophytes or lipping present without porosity. Porosity was evident without lipping or osteophytes in some cases). Eburnation of the joint surface or pathologic exposure of trabecular bone was also scored as arthropathy. Arthropathy was considered present in a joint when any part of the joint was affected (e.g., distal femur and proximal tibia were both scored as "knee"). The category ankle/wrist includes distal radius, ulna, tibia, fibula, and any carpals or tarsals. The category hand/foot includes metacarpals, metatarsals, and all phalanges.

Of the 20 adults included in this analysis, 18 exhibit arthropathy in at least one joint (8 females, 10 males). See Tables 2.7, 2.8, and 2.9 for full results. In this small sample, the presence/absence of at least one joint being affected by arthropathy was not mediated by age (Mann–Whitney $U = 4.00$, $p = 0.08$), though there may be a moderate correlation between age and the number of joints/regions affected by arthropathy (Fig. 2.5). Linear regression analysis in ARC (Cook and Weisberg 1999) indicates that $R^2 = 0.3340$, with $p = 0.0076$. The case for an approximately linear relationship between number of joints/regions affected by arthropathy and age is strengthened by an examination of the two cases with the greatest number of joints/regions affected by arthropathy: Bwavi Male and Goblin (23 and 17 joints/regions affected, respectively). Goblin was nearly 40 years old when he died, and Bwavi Male is estimated to have reached approximately the same age. I used Dixon's test for outliers (Dixon 1953) to determine whether either of these chimpanzees might be considered a mathematical outlier. With $n = 20$, the critical value for this test is 0.450 (Ibid.). If the test statistic r exceeds the critical value, the case is likely to be a mathematical outlier. While Goblin should not be considered an outlier ($r = 0.411$), Bwavi Male may indeed be an exceptional case ($r = 0.565$). Given the extreme morphological changes described in Chap. 1 for this adult male chimpanzee, Bwavi Male likely suffered from an extreme condition that rarely affects wild chimpanzees. Removing his case, however, does not significantly strengthen the relationship between age and the number of joints/regions affected by arthropathy ($R^2 = 0.3404$, $p = 0.0087$ in the reduced sample, versus $R^2 = 0.3340$, with $p = 0.0076$ in the full sample).

Table 2.7 The presence/absence of arthropathy in adult chimpanzee axial skeletons and acromio-clavicular joint

Specimen	Sex	Age	C vert	T vert	L vert	TMJ L	TMJ R	Ribs L	Ribs R	Sacroiliac L	Acromio-clavicular L	Sterno-clavicular L	Sacro-iliac R	Acromio-clavicular R	Sterno-clavicular R
Sherehe	F	15.78	0	0	0	0	0	0	0	0	0	0	0	0	0
Gilka	F	18.87	0	0	0	0	0	0	0	0	0	0	0	0	0
Yolanda	F	24	0	0	0	0	0	1	1	0	0	0	0	0	0
Kidevu	F	25.59	0	0	0	0	0	0	0	0	0	0	0	0	0
Charlie	M	25.87	0	0	0	0	0	0	0	0	0	0	0	0	0
Pallas	F	27.2	0	0	0	0	0	0	0	0	0	0	0	0	0
Madam Bee	M	28.22	0	0	0	0	0	0	0	0	0	0	0	0	0
Vincent	M	28.47	0	0	0	0	0	1	1	0	1	0	0	1	0
Tumaini	M	30	0	0	0	0	0	0	0	0	0	0	0	0	0
Passion	F	30.61	0	0	0	0	0	1	1	0	0	0	0	0	0
Jomeo	M	30.87	0	0	0	0	0	0	0	0	0	0	0	0	0
Miff	F	30.9	0	0	0	1	1	0	0	0	0	0	0	0	0
Atlas	M	31.3	1	1	0	0	0	0	0	0	1	0	0	0	0
Echo	F	34	1	1	0	0	0	0	0	0	0	0	0	0	0
Melissa	F	36.31	0	0	0	0	0	0	0	0	0	0	0	0	0
Hugo	M	38.58	0	0	0	1	1	1	1	0	0	0	0	0	0
Goblin	M	39.96	1	1	1	1	1	1	1	1	1	0	0	1	0
Bwavi Male	M	40	1	1	1	1	1	1	1	1	1	1	1	1	1
Patti	F	44.25	0	1	0	0	0	1	1	0	0	0	0	0	0
Old Female	F	50	0	1	0	0	0	0	0	0	0	0	0	0	0

0 = no change or mild porosity, 1 = lipping/osteophytes and porosity OR eburnation

Table 2.8 The presence/absence of arthropathy in adult chimpanzee appendicular skeletons

Specimen	Sex	Age	Ankle/wrist	Hand/foot	Shoulder L	Elbow L	Hip L	Knee L	Shoulder R	Elbow R	Hip R	Knee R
Sherehe	F	15.78	0	0	0	0	0	0	0	0	0	0
Gilka	F	18.87	1	0	0	0	0	0	0	1	0	0
Yolanda	F	24	1	0	0	0	1	1	0	0	0	0
Kidevu	F	25.59	0	0	0	1	0	0	0	0	0	0
Charlie	M	25.87	0	0	0	0	0	0	1	0	0	0
Pallas	F	27.2	0	0	0	0	0	0	0	0	0	0
Madam Bee	M	28.22	0	1	0	1	0	0	0	1	0	0
Vincent	M	28.47	0	1	0	0	0	0	0	0	0	0
Tumaini	M	30	1	0	0	0	0	0	0	0	0	0
Passion	F	30.61	1	0	0	1	1	0	0	1	1	0
Jomeo	M	30.87	0	0	0	1	0	0	1	1	1	0
Miff	F	30.9	0	1	0	1	0	0	1	1	1	0
Atlas	M	31.3	0	1	1	1	1	0	1	1	1	0
Echo	F	34	0	0	0	1	0	1	0	1	0	1
Melissa	F	36.31	0	0	0	0	0	0	0	0	0	0
Hugo	M	38.58	1	1	0	1	0	1	0	1	0	1
Goblin	M	39.96	1	1	1	1	0	1	0	1	0	1
Bwavi Male	M	40	1	1	1	1	1	1	1	1	1	1
Patti	F	44.25	1	1	0	1	1	1	0	0	0	0
Old Female	F	50	0	0	1	0	1	1	1	1	0	1

0 = no change or mild porosity, 1 = lipping/osteophytes and porosity OR eburnation

Table 2.9 Summary arthropathy data

Specimen	Sex	Age	Any location	Forelimb	Hindlimb	Total	Related to trauma?
Sherehe	F	15.78	0	0	0	0	No
Gilka	F	18.87	1	1	0	2	Other path
Yolanda	F	24	1	0	1	5	Maybe
Kidevu	F	25.59	1	1	0	1	No
Charlie	M	25.87	1	1	0	1	No
Pallas	F	27.2	0	0	0	0	No
Madam Bee	M	28.22	1	1	0	3	Yes
Vincent	M	28.47	1	1	1	5	Yes
Tumaini	M	30	1	0	1	1	Yes
Passion	F	30.61	1	1	0	7	Yes
Jomeo	M	30.87	1	1	1	4	Yes
Miff	F	30.9	1	1	1	7	Yes
Atlas	M	31.3	1	1	1	10	Yes
Echo	F	34	1	1	1	6	Yes
Melissa	F	36.31	1	1	0	0	No
Hugo	M	38.58	1	1	1	10	Yes
Goblin	M	39.96	1	1	1	17	Yes
Bwavi Male	M	40	1	1	1	23	Maybe
Patti	F	44.25	1	1	1	8	Maybe
Old Female	F	50	1	1	1	7	Yes

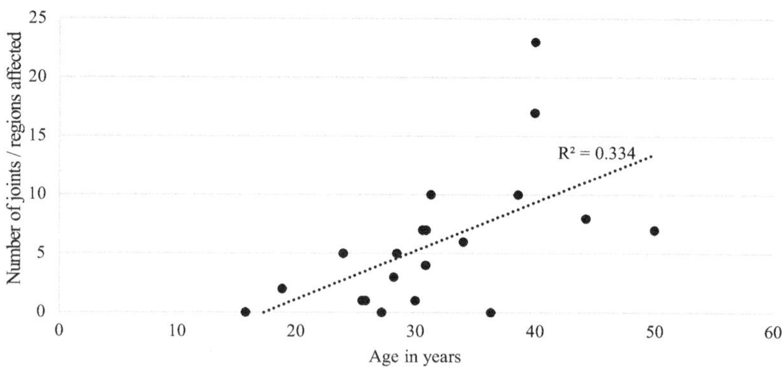

Fig. 2.5 Number of regions/joints affected by arthropathy versus age

An additional factor to consider is that Bwavi Male's age is truly an estimate. I estimated his age as 40 years, as his teeth resemble Patti's wear stage (she was 44 when she died), and are in much better condition than Flo's teeth (who is estimated to have been around 50 when she died). The age estimate for Bwavi Male may not be accurate, however. Hugo, for example, exhibits much greater antemortem tooth loss, and he is estimated to have been 38 when he died. Goblin, aged 40, exhibits a more advanced wear stage than Bwavi Male. If this male chimpanzee were actually around 50 when he died, his case would not seem to deviate from the observed trend. Conversely, if he were younger than Goblin, Bwavi male would represent a more extreme outlier.

Old Female is another animal who noticeably diverges from the trend line (coordinates 50, 7). This is another case where the age estimate may be inaccurate. Her degree of dental wear and antemortem tooth loss is comparable to Flo, so I estimated that Old Female was also around 50 years old when she died. It is worth emphasizing that Flo's age itself is an estimate. It may be valuable to re-assess such age estimates in future after careful study of tooth wear, attrition, and antemortem tooth loss in older chimpanzees of known age. Most of the old chimpanzees currently included in the skeletal sample have age estimates rather than known ages; Goblin and Patti are notable exceptions.

Further, age is not the only factor likely to influence arthropathy. With a larger skeletal sample and accompanying long-term behavioral and demographic data, it might be possible to further explore intersections between age, sex, and other conditions. The opportunity to test for different trends in males versus females would be particularly compelling, since arthropathy is often considered a sequel to trauma in wild animals, and sex differences in the frequency of traumatic lesions across body regions are evident in this sample (see below).

2.7 Dental Lesions

Because the dentition is often analyzed separately from the rest of the skeleton, I present a separate section on dental lesions. The remainder of this chapter integrates dental, craniofacial, and post-cranial lesions in order to approach an analysis of the entire animal.

To examine dental lesions, I evaluated 23 adult chimpanzees (10 males, 13 females) for evidence of antemortem tooth loss, osteoarthritis of the temporomandibular joint, periapical abscesses, other periodontal disease (usually alveolar recession), dental caries, pulp cavity exposure (due to either breakage or wear), chipped teeth, other dental trauma, and supernumerary teeth/other congenital anomalies. Frequencies of each type of lesion are depicted in Fig. 2.6. I also tested for male–female differences in lesion rates using Fisher's exact test in SPSS version 24.0 (IBMCorp 2016). Dental caries may be more frequent in males relative to females in this sample, but there were no other sex differences in lesion frequency (Table 2.10). The small sample size calls for caution in interpreting these results. The single case of "other trauma" to the dentition is Tumaini, whose anterior teeth are calcined. Tumaini was a poaching victim and his body was partially burned (see Chap. 1).

I tested for a relationship between age and two kinds of lesions often associated with degenerative disease/senescence: antemortem tooth loss and osteoarthritis of

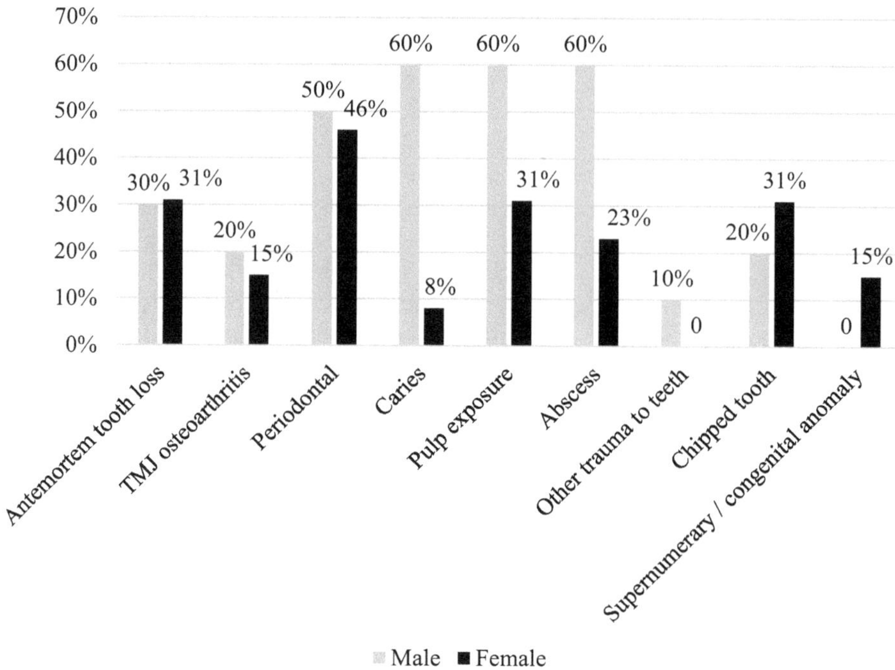

Fig. 2.6 Frequencies of dental and related lesions

the TMJ. Both of these relationships only approach significance, though the small sample sizes ($n = 23$ for tooth loss, $n = 20$ for TMJ osteoarthritis) are almost certainly a barrier to detecting real relationships between age and these lesion types.

There is a moderate correlation between age and the number of adult teeth lost antemortem ($R^2 = 0.305$, Fig. 2.7), but the relationship only approaches significance ($p = 0.053$). Additional data may provide evidence for an exponential rather than a linear relationship between antemortem tooth loss and age. If that is the case, it would match previous conclusions about antemortem tooth loss, tooth wear, and other dental pathologies increasing with age that are based on a smaller sub-set of the currently available sample (Morbeck et al. 2002).

Similarly, the difference between the ages of chimpanzees with and without osteoarthritis of the TMJ only approaches significance ($U = 11.00$, $p = 0.050$). A larger sample might provide more convincing evidence that chimpanzees with TMJ osteoarthritis tend to be older than chimpanzees who do not exhibit this lesion (*mean rank* for animals with TMJ osteoarthritis = 15.75, without = 9.19).

Table 2.10 Fisher's exact test results for male–female differences in the frequencies of dental and related lesions

		Antemortem tooth loss	TMJ osteoarthritis	Periodontal	Caries	Broken tooth	Abscess	Other trauma to teeth	Chipped tooth	Supernumerary/ congenital anomaly
Male	10	3 (30%)	2 (20%)	5 (50%)	6 (60%)	6 (60%)	6 (60%)	1 (10%)	2 (20%)	0
Female	13	4 (31%)	2 (15%)	6 (46%)	1 (8%)	4 (31%)	3 (23%)	0	4 (31%)	2 (15%)
Total	23	7 (30%)	4 (17%)	11 (48%)	7 (30%)	10 (43%)	9 (39%)	1 (4%)	6 (26%)	2 (9%)
p-value for Fisher's exact test for M–F differences		1.00	0.281	0.66	**0.019**	0.222	0.221	Not calculated	0.604	Not calculated

Bold indicates a statistical difference at $\alpha = 0.05$

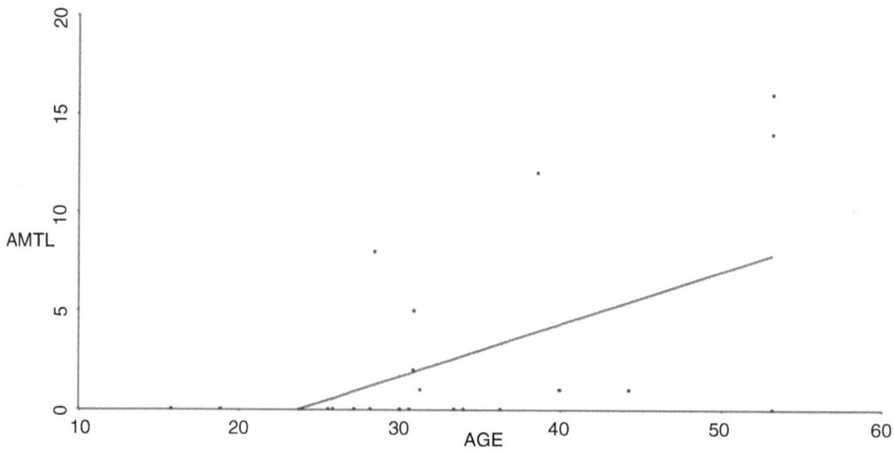

Fig. 2.7 Possible relationship between age and number of teeth lost antemortem (AMTL). Graph generated using ARC for regression analysis (Cook and Weisberg,1999). $R^2 = 0.305$, $p = 0.053$

2.8 Are There Sex Differences in Rates of Trauma? Are There Sex Differences in Location/Pattern of Trauma?

Mann–Whitney tests provide no evidence for sex differences in overall rate of trauma ($U = 132.5$, $p = 0.508$), pathology ($U = 143.0$, $p = 0.765$), or both types of lesion combined ($U = 121.0$, $p = 0.304$) for the full sample of 36 complete chimpanzee skeletons of known sex. There may be sex differences in trauma rates between different regions of the body. I examine traumatic lesions by region in two ways:

1. Evaluating the proportion of total traumatic lesions that occurred in each body region (proportion of lesions).
2. Evaluating the proportion of the observed skeletal elements (by region) affected by traumatic lesions (regional trauma rate).

2.9 Analysis of Proportion of Traumatic Lesions

Figure 2.8 shows which body regions were most commonly affected by trauma for males and females. The distal extremity was the most commonly affected region for males (36% of traumatic lesions to males occurred in this region), while the trunk was the most commonly affected region for females (42% of observed traumatic lesions in females).

The proportion of traumatic lesions by region in males versus females (Fig. 2.9) does not differ for head traumata, but the rate for other regions may differ significantly (z-score test results reported in Table 2.11). The proportion of traumatic

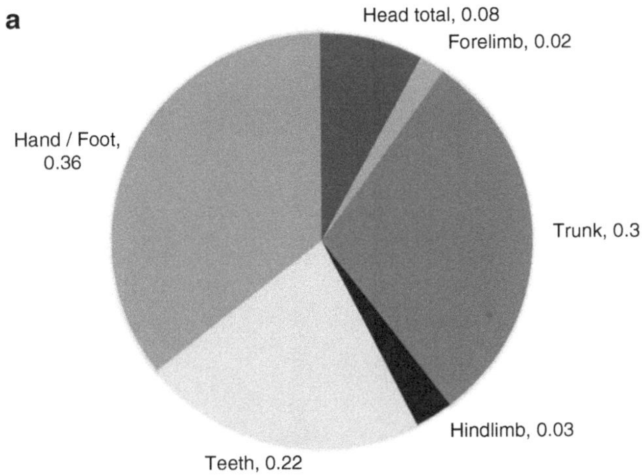

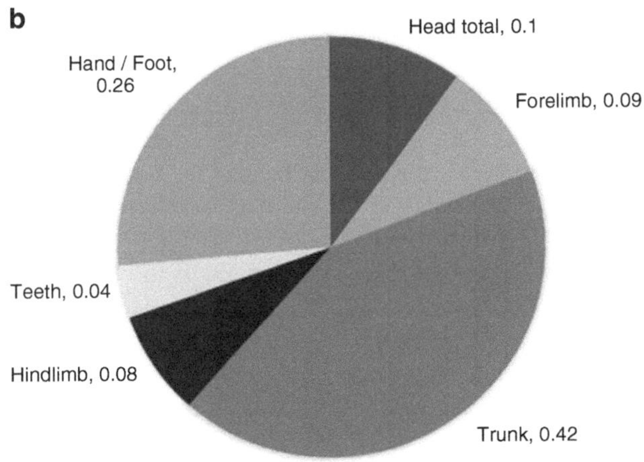

Fig. 2.8 (**a**) Percentage of total traumatic lesions occurring in each body region for males. (**b**) Percentage of total traumatic lesions occurring in each body region for females

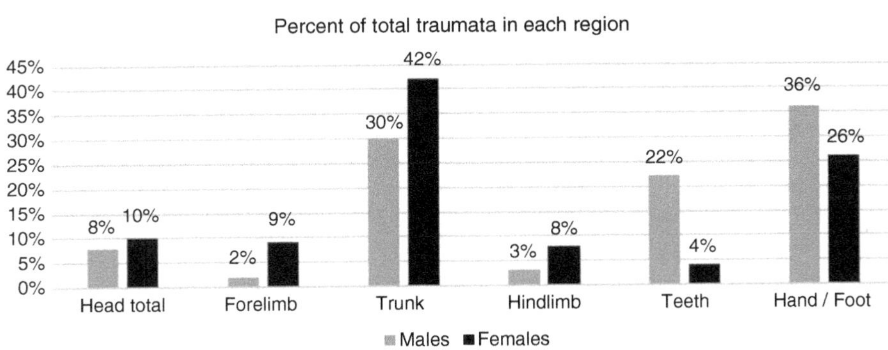

Fig. 2.9 Contrasting regional trauma rates for males versus females

Table 2.11 Two-tailed z-score tests examining whether the percentage of traumatic lesions by region differs for males versus females

Region	z	p
Head	−0.67	0.5
Forelimb	−3.5	0.0004[a]
Trunk	**−3**	**0.003**
Hindlimb	−2.5	0.0124[a]
Teeth	4.5	~0[a]
Hand/foot	**2.5**	**0.0124**

$\alpha = 0.05$. Bold indicates a statistical difference
[a]Indicates that the sample size is too small for the test to be statistically valid

lesions affecting the limbs and trunk may be higher in females, while the proportion of traumatic lesions affecting the distal limbs and teeth may be higher in males. Limited sample sizes mean that some of the tests are not statistically valid. Results should therefore be interpreted with caution and regarded as a starting point for further long-term research. This quantitative exploration nevertheless provides insight into demographic trends emerging from this sample.

While the proportion of injuries sustained to the head was the same for both sexes ($p = 0.5$, see Table 2.11), the ratio of head: fore limb traumata is higher in males (4.0) compared to females (1.1), which is consistent with previous work (Carter et al. 2008). In addition, the distribution of cranial injuries differs between males and females, with 100% of observed injuries to the head (excluding teeth) occurring on the face for males, while only 8% of observed cranial injuries occurred on the face for females (face versus vault regions defined by Novak and Hatch (2009), who did not examine dental injuries). In other words, Gombe chimpanzee females in this sample experienced more trauma to the cranial vault while males experienced more trauma to the face, which is consistent with previous results on sex differences in cranial lesion distribution (Novak and Hatch 2009; Jurmain and Kilgore 1998).

2.10 Sex Differences in Trauma Rate

Similar patterns emerge in terms of the distribution of traumatic lesions across observed skeletal elements (trauma rate) (Fig. 2.10), though with some relevant differences compared to the proportion of lesions analysis. Of the observed skeletal elements, the head was the most commonly affected region, with 29% of the observed male skulls and 50% of the observed female skulls exhibiting traumatic lesions. While this only approached being significantly different between sexes ($p = 0.08$, see Table 2.12), other sex differences in the rate of traumatic lesions are notable. In contrast to the proportion of lesions analysis, there are no male–female

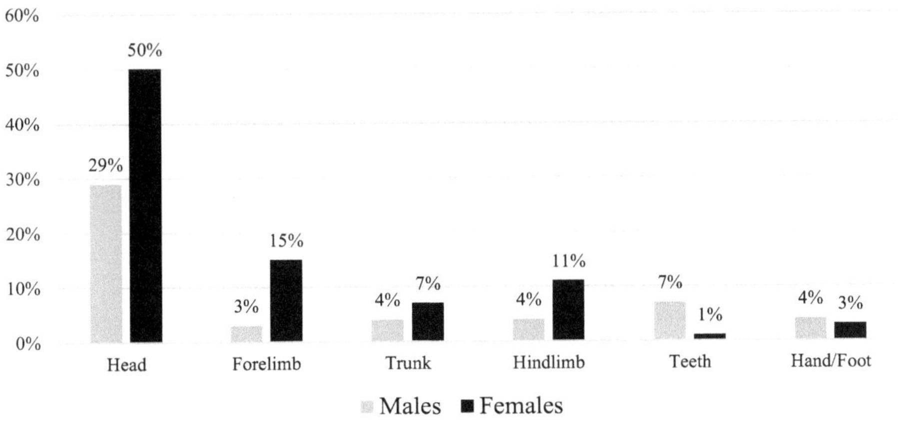

Fig. 2.10 Percentage of observed elements affected by trauma (trauma rate)

Table 2.12 Two-tailed z-score tests examining whether the percentage of observed skeletal elements (by region) differs between males and females

Region	z	p
Head	−1.75	0.08
Forelimb	−1.09	0.93[a]
Trunk	**−3.33**	**0.0008**
Hindlimb	−2.19	0.03[a]
Teeth	4.62	~0[a]
Hand/foot	1.25	0.21

$\alpha = 0.05$. Bold text indicates a statistical difference

[a]Indicates that the sample size is too small for the test to be statistically valid

differences in trauma rate in the trunk, hindlimb, and teeth. Females may have a higher trauma rate to the trunk and hindlimb regions, while males likely have a higher trauma rate to the teeth. Similar concerns about sample size also urge caution in the interpretation of these results (see Table 2.12).

2.11 Do Rates of Trauma Differ Between Traumatic Versus Non-Traumatic Causes of Death?

The percentage of observed skeletal elements with traumatic lesions differs between animals (male and female) with traumatic versus non-traumatic causes of death ($U = 47.00$, $p = 0.002$). This is also true when examining females only ($U = 4.00$, $p = 0.003$), but not males only ($U = 22.00$, $p = 0.212$). This may be due to some influential cases (female infanticide victims) discussed in the next section examining the relationship between age and traumatic lesions.

2.12 Do Trauma Rates Increase with Age?

In the sample of 36 chimpanzees with known age at death (or estimated age considered to be accurate, see Table 1.2), there is no relationship between age and the percentage of observed bones affected by traumatic lesions ($R^2 = 0.0378, p = 0.2556$). For all of the analyses discussed in this section, age and the percentage of elements affected by trauma were log-transformed because variables can differ by an order of magnitude, and then regressed using ARC statistical software (Cook and Weisberg 1999). These variables are still referred to by their untransformed names (age, trauma rate, or the percentage of observed bones affected by traumatic lesions) for ease of reading.

Three influential cases are evident in Fig. 2.11. The circled data points are Andromeda and Rejea, while the data point enclosed in a box represents Vincent. All three of these cases are animals killed by conspecifics, whose percentage of skeletal elements affected by trauma is an order of magnitude higher than the other animals in the sample.

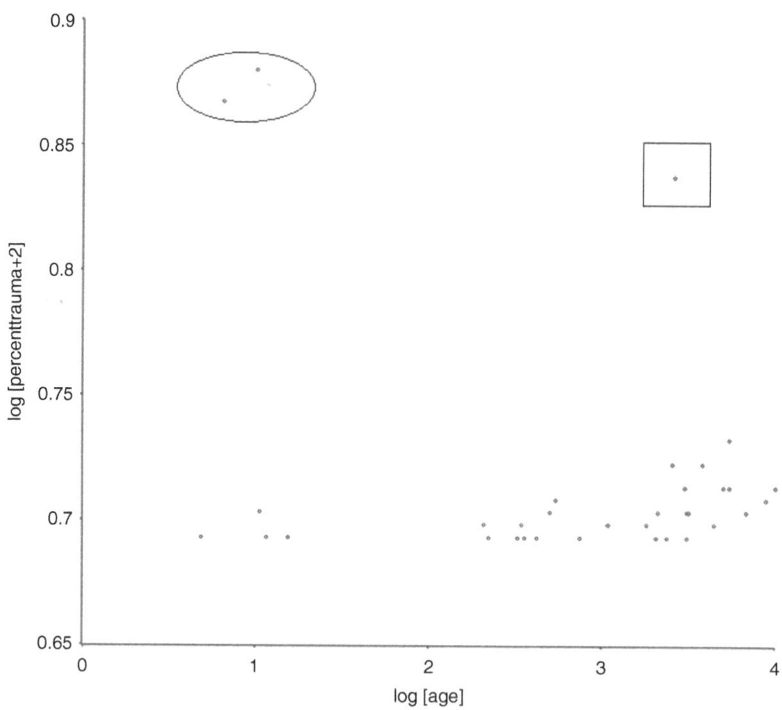

Fig. 2.11 Linear regression demonstrating no relationship between age and the rate of traumatic lesions. Circled data points are Andromeda and Rejea, data point with a square is Vincent, all influential cases. Figure generated using ARC statistical software (Cook and Weisberg 1999)

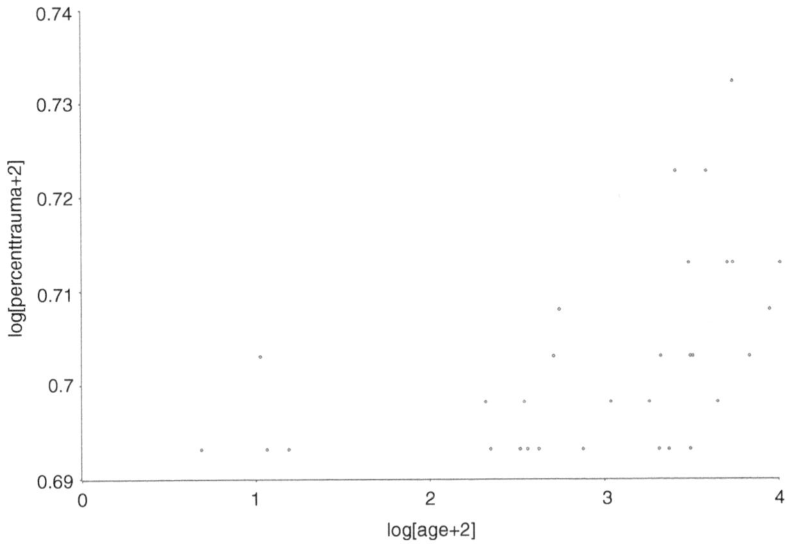

Fig. 2.12 Linear regression with influential cases removed. Figure generated using ARC statistical software (Cook and Weisberg 1999)

If those three influential cases are removed, a moderate relationship between age and the percentage of bones affected by trauma emerges ($R^2 = 0.2845$, $p = 0.0014$), which is consistent with previous work (Jurmain 1989, 1997). Figure 2.12 illustrates that the relationship may not, however, be linear. Furthermore, removing these three influential cases is not a biologically appropriate way to address this data set, since conspecific aggression is a significant source of mortality for the Gombe chimpanzees (Williams et al. 2008).

Given that the rates of trauma differ for chimpanzees with traumatic versus non-traumatic deaths (see above), I also analyzed these two groups separately. For chimpanzees with traumatic causes of death ($n = 11$), there is a significant negative correlation between percentage of bones affected by trauma and age ($R^2 = 0.4618$, $p = 0.0214$, see Fig. 2.13), which is driven by the two infanticide cases (Andromeda and Rejea).

In the sample of chimpanzees with non-traumatic causes of death (illness) ($n = 25$), there is a significant, though moderate, positive correlation between age and the percentage of bones affected by trauma ($R^2 = 0.2627$, $p = 0.0088$). Figure 2.14 illustrates that this relationship may not be linear, and Gamma regression suggests that this relationship may be exponential ($\alpha = 1$, $p = 0.0039$).

To further explore the relationship between age and trauma rate, I also examined a sub-sample including adults only (all causes of death, $n = 20$). No significant relationship between age and trauma rate seems to exist ($R^2 = 0.0046$, $p = 0.7769$), though again, this seems to be affected by an influential case: Vincent (Fig. 2.15). Indeed, Vincent can be considered a mathematical outlier according to Dixon's test for outliers ($n = 20$ adults, $r = 0.807$, critical value = 0.450) (Dixon 1953; Kanji 1999).

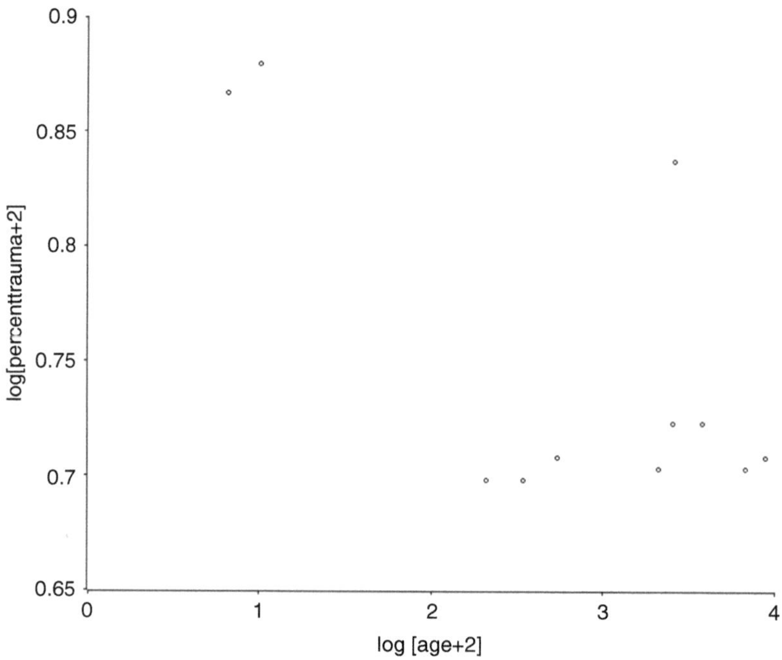

Fig. 2.13 Negative correlation between age and percent of bones affected by trauma. Note the influence of the two infanticide cases, which appear farthest to the left on the x-axis. Figure generated using ARC statistical software (Cook and Weisberg 1999)

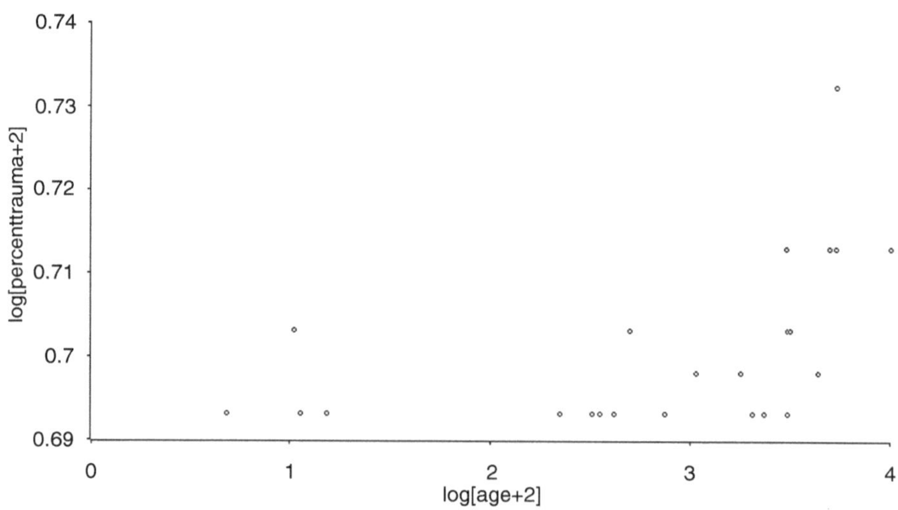

Fig. 2.14 Exponential relationship demonstrated between age and the percentage of bones affected by trauma in a sub-sample including only chimpanzees with non-traumatic causes of death. Figure generated using ARC statistical software (Cook and Weisberg 1999)

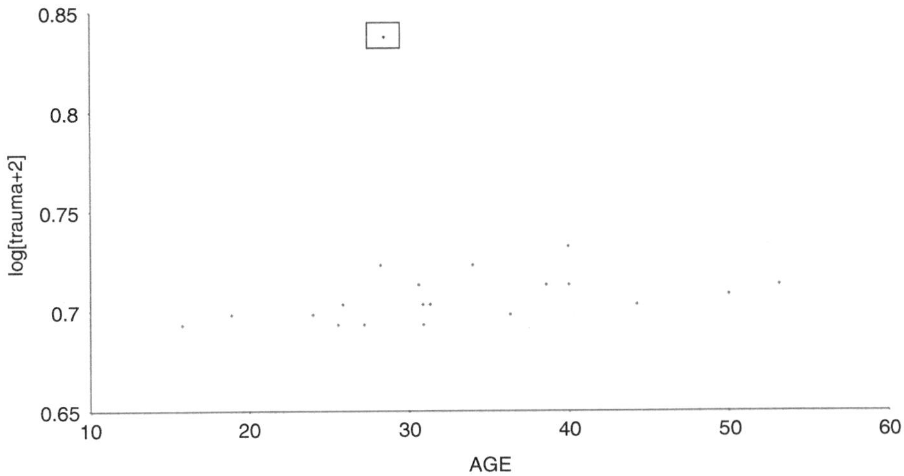

Fig. 2.15 No relationship between age and trauma rate in a sub-sample including adults only. The influential case enclosed with a box is Vincent, who was killed by conspecifics. Figure generated using ARC statistical software (Cook and Weisberg 1999)

Vincent was, however, killed by two other adult males in his home community (Mitumba); conspecific aggression is a significant source of mortality for chimpanzees, and is the second most-common cause of death for males from Kasekela (Williams et al. 2008). While, from a statistical standpoint, a clearer relationship between age and trauma rate might emerge if Vincent's case were removed from the analysis, eliminating his important case from the sample would have no behavioral or biological basis.

Skeletal trauma rate in this sample of the Gombe chimpanzees seems to increase exponentially with age, though this relationship is not always consistent in cases of traumatic death, when skeletal trauma rates may be an order of magnitude higher than for animals who died of non-traumatic causes. It is also worth noting that skeletal signs of traumatic death are extremely variable, with animals like Rejea, Andromeda, and Vincent exhibiting extremely high rates of trauma, and others showing few skeletal signs of their cause of death. Patti is a notable example of the latter. Only 2% of Patti's bones have traumatic lesions ($n = 5$ lesions), and only one lesion can be attributed to her cause of death, as all the others are well healed. Patti was killed by male chimpanzees in 2005, in a manner similar to Vincent, and researchers noticed extensive soft tissue damage to Patti's body at the time of necropsy (see case studies in Chap. 1). In contrast, 30% of Vincent's bones exhibit skeletal evidence of trauma, and approximately half of those lesions were perimortem (see Table 2.1a, 2.1b, and 2.1c).

2.13 Are There Sex Differences in Rates of Pathology? Are There Sex Differences in Location/Pattern of Pathology?

As in the analysis of traumatic lesions, there is no detectable difference in the rate of pathologic lesions between males and females ($U = 143.00$, $p = 0.765$), where the rate of pathologic lesions is calculated as the proportion of observed bones affected by pathologic lesions. As with trauma, I examine pathologic lesions by region in two ways:

1. Evaluating the proportion of total pathologic lesions that occurred in each body region (proportion of lesions).
2. Evaluating the proportion of the observed skeletal elements (by region) affected by pathologic lesions (pathology rate).

2.14 Analysis of Proportion of Pathologic Lesions

In both males and females, the distal extremities make up the highest proportion of lesions (Fig. 2.16). While the proportion of total pathologic lesions was exactly the same between males and females for the teeth as well as for unidentified elements (Fig. 2.17), some sex differences by region are evident (Table 2.13). Males have a higher proportion of forelimb and trunk lesions.

2.15 Sex Differences in Pathology Rate

The unidentified elements affected by pathology are all hard tissue of irregular morphology that likely resulted from heterotopic ossification (e.g., in Melissa and Goblin, see Chap. 1). Therefore, 100% of the observed unidentified bones are affected by pathology, and this is true for both sexes. The pathology rates in the teeth and distal limbs do not differ between males and females (Table 2.14), though differences in other body regions are notable (Fig. 2.18). While the pathology rate of the forelimb is higher in females compared to males, the rate is higher in the trunk and hindlimb for males. Rate of pathological lesions in the head may be higher for males, but the sample size is too small for this test to be statistically valid.

2.16 Do Pathology Rates Increase with Age?

The relationship between pathology rate (number of observed bones with pathologic lesions) and age is more straightforward than the relationship between trauma rate and age. There is a significant, positive relationship between age and pathology

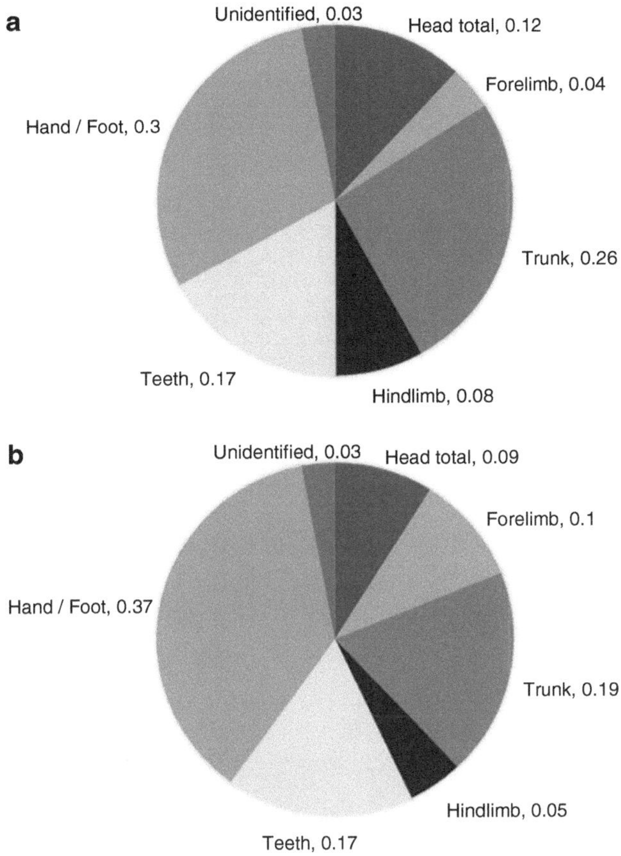

Fig. 2.16 (**a**) Proportion of the total number of pathologies (n = 333) by region for males. (**b**) Proportion of the total number of pathologies (n = 230) by region for females

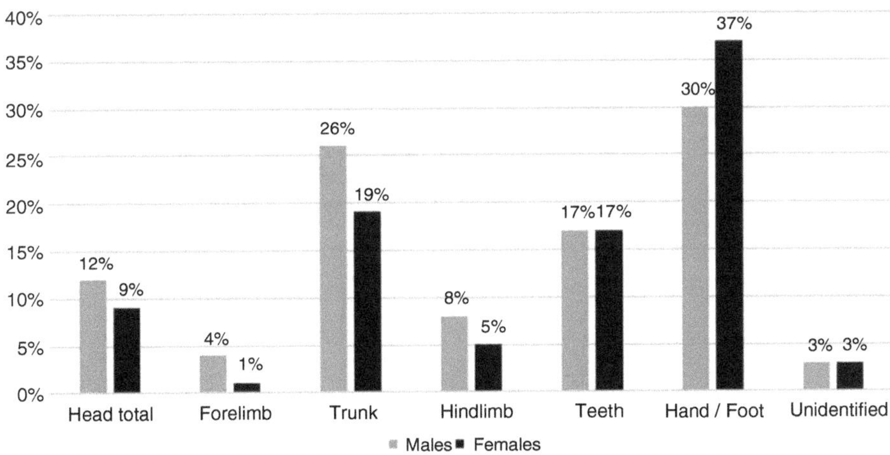

Fig. 2.17 Contrasting the proportion of pathologic lesions by region between males and females

Table 2.13 Two-tailed z-score tests examining whether the regional pathology rate differs for males versus females

Region	z	p
Head	1	0.32
Forelimb	**−3.00**	**0.003**
Trunk	**3.5**	**0.0004**
Hindlimb	0.375	0.7
Teeth	0	1
Hand/foot	−1.75	0.08
Unidentified	0	1

$\alpha = 0.05$. Bold indicates a statistical difference

Table 2.14 Two-tailed z-score tests examining whether the percentage of observed skeletal elements affected by pathology (by region) differs between males and females

Region	z	p
Head	3.22	0.001[a]
Forelimb	**−3.25**	**0.001**
Trunk	**4**	**~0**
Hindlimb	**2.25**	**0.02**
Teeth	1.5	0.13
Hand/foot	0	~1

$\alpha = 0.05$. Bold text indicates a statistical difference
[a]Indicates that the sample size is too small for the test to be statistically valid

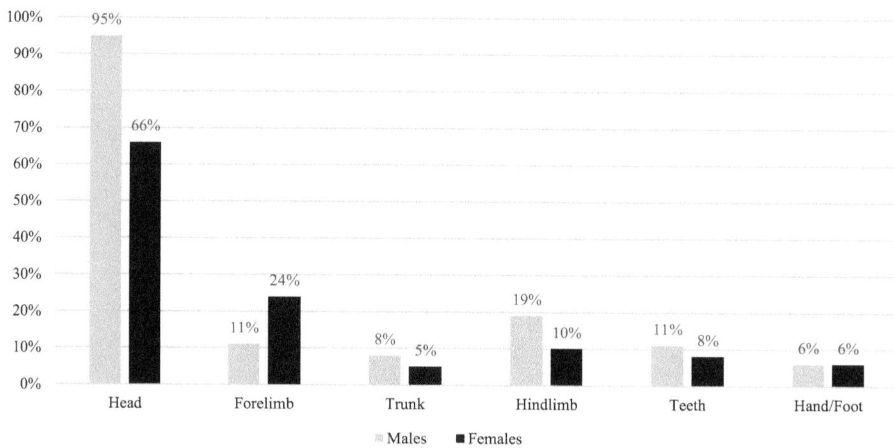

Fig. 2.18 Regional pathology rate, sex comparisons

rate (Fig. 2.19, $R^2 = 0.5034$, $p = 0$). Untransformed results are shown in the graph because the regression analysis using log-transformed data was extremely similar ($R^2 = 0.5009$, $p = 0$). The variation in residuals between data points is normally distributed, and reflects the variation in selective pressures experienced by individual chimpanzees.

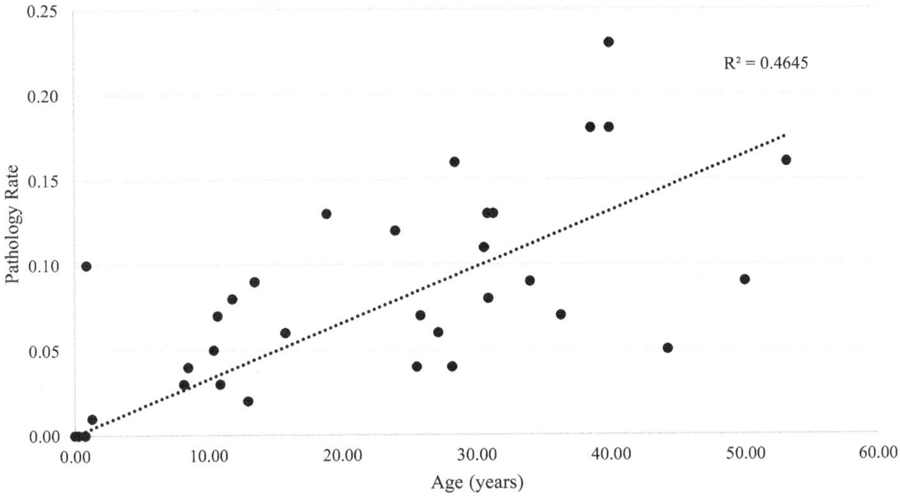

Fig. 2.19 Pathology rate increases with age

2.17 Do Rates of Skeletal Lesions Differ Between Dominance Rank Categories?

Trauma rate and pathology rate are calculated as the percentage of observed bones affected by lesions attributed either to traumatic or other pathologic causes. Kruskal–Wallis tests determined whether the three rank categories used in this study (1 = high ranking, 2 = middle ranking, 3 = low ranking, see Chap. 1) affect the rate of skeletal lesions. Krukall–Wallis is a "rank-sum" test wherein each animal is assigned an ordinal rank within the sample based on their lesion rate. This statistical ordinal rank is then used to calculate the "mean rank" for each dominance rank category (Kruskal and Wallis 1952). The *mean rank* is therefore calculated using the Kruskal–Wallis method in SPSS (IBMCorp 2016), while dominance rank is based on behavioral data of the chimpanzees from Gombe while they were alive (see Table 1.3). The term *mean rank* will appear in italics throughout this chapter to highlight that it refers to a calculated statistic rather than an animal's dominance rank. For example, Vincent's highest achieved dominance rank during his lifetime was high ranking, and his assigned ordinal rank for trauma rate is 34th (he has the third highest trauma rate, see Table 2.1a, 2.1b, and 2.1c). The ordinal ranks assigned to each high-ranking animal are used to calculate the *mean rank* for all high-ranking animals. I discuss the calculated *mean rank* for groups of animals in selected dominance rank categories, as relevant.

I evaluated three different measures of dominance rank for each animal whenever possible: the individual's highest achieved rank, mother's rank at time of the individual's birth, (mother's rank), and the individual's rank at the time of death

Table 2.15 Results from Kruskal–Wallis tests examining the possible relationship between dominance rank categories and trauma rate

Samples: Trauma rate	n	χ^2	df	p
All chimps, highest rank	30	4.485	2	0.106
All chimps, mother's rank	16	3.599	2	0.165
All chimps, death rank	31	3.987	2	0.136
Males, highest rank	**17**	**7.067**	**1**	**0.008 ($r = 0.645$)**
Males, mother's rank	12	2.489	2	0.288
Males, death rank	18	2.382	1	0.123
Females, highest rank	13	0.519	2	0.771
Females, mother's rank	4	0.889	1	0.346
Females, death rank	13	1.703	2	0.427

Bold text indicates a significant relationship. $\alpha = 0.05$

(death rank). See Table 1.3 for dominance rank data on individual chimpanzees. I evaluated the relationship between dominance rank and skeletal lesions for all chimpanzees and for males and females separately.

In the complete sample including males and females, rates of trauma do not differ between dominance rank categories, whether the rank assessment is based on highest achieved dominance rank, mother's rank, or rank at the time of death (Table 2.15). Trauma rate, however, is mediated by highest achieved dominance rank for male chimpanzees ($\chi^2 = 7.067$, df = 1, $p = 0.008$). The *mean rank* for high-ranking males is 13.80, while for low-ranking males, the *mean rank* is 7.0 (there were no males whose highest achieved rank was middle rank), indicating that trauma rates for high-ranking males are likely higher. I also calculated effect size to determine how much of the variation in trauma rate can be attributed to dominance rank, according to procedures outlined by Rosenthal and DiMatteo (2001) to calculate r. The effect size of dominance rank on male trauma rate is $r = 0.645$, which is considered a large effect size (Cohen 1992).

In Chap. 1, I hypothesized that high-ranking males sustain more skeletal injuries than other demographic categories due to increased agonistic encounters. There were no detectable differences in trauma rates between rank categories for female chimpanzees or in the combined sex sample. Thus, this data set provides no support for the second hypothesis that low-ranking females would exhibit higher trauma rates than conspecifics from other demographic categories. Later in this chapter, I address whether variation between rank categories is driven by age.

Pathology rate differs between dominance rank categories for highest achieved rank in the whole sample ($\chi^2 = 10.460$, df = 2, $p = 0.005$) as well as in a sub-sample of males only ($\chi^2 = 7.861$, df = 1, $p = 0.005$). There were no statistical differences in pathology rate between other measures of dominance rank or in a sub-sample of females only (Table 2.16). In the skeletal pathology analysis, there are similar patterns relative to the trauma analysis: high-ranking males seem to have a higher pathology rate than males of low dominance rank (*mean rank* = 14.30 for high-ranking males, *mean rank* = 6.79 for low-ranking males).

Table 2.16 Results from Kruskal–Wallis tests examining the possible relationship between dominance rank categories and pathology rate

Samples: Pathology rate	n	χ^2	df	p
All chimps, highest rank	**30**	**10.46**	**2**	**0.005 ($r = 0.590$)**
All chimps, mother's rank	16	0.035	2	0.982
All chimps, death rank	31	4.882	2	0.0878
Males, highest rank	**17**	**7.861**	**1**	**0.005 ($r = 0.680$)**
Males, mother's rank	12	0.022	2	0.989
Males, death rank	18	2.389	1	0.122
Females, highest rank	13	2.387	2	0.303
Females, mother's rank	4	0.2	1	0.655
Females, death rank	13	2.842	2	0.242

Bold text indicates a significant relationship. $\alpha = 0.05$

The difference detected in pathology rate between chimpanzees of different highest achieved ranks in the combined sex sample is likely driven by the differences detected in the sub-sample including only males. The *mean rank* for high-ranking chimpanzees is 22.15, and for middle-ranking it is 18.50, but for low-ranking animals, the *mean rank* is 11.06. This may indicate a greater difference between high- and low-ranking animals compared to high- and middle-ranking animals, which may be related to the fact that this sample includes no males whose highest achieved dominance rank is middle rank. The effects size ($r = 0.680$ (Rosenthal and DiMatteo 2001)) for the influence of highest achieved dominance rank on pathology rate for male chimpanzees is large (Cohen 1992), similar to the finding for trauma rate.

2.18 Does Age Account for the Effect Size of the Influence of Rank on Skeletal Lesions?

The previous section indicates that dominance rank may influence skeletal lesion rate in the Gombe chimpanzees, suggesting that high-ranking males experience a higher rate of skeletal lesions. Based on this sample as well as previous skeletal analyses of chimpanzees, age likely also influences skeletal lesion rate.

Age also mediates dominance rank to a certain extent, given that all sub-adult chimpanzees are low ranking. I therefore used Kruskal–Wallis tests to explore a possible relationship between dominance rank and age, and used the Rosenthal and DiMatteo (2001) method to calculate effects sizes (Table 2.17). Both highest achieved rank and individual's rank at the time of death are mediated by age, with large (Cohen 1992) effects sizes. That mother's rank does not seem to differ by age in this sample may be influenced by the smaller available sample of chimpanzees for whom this type of data is available. Of particular note is that the effects sizes for

Table 2.17 Results for Kruskal–Wallis tests examining the possible relationship between dominance rank categories and age

Samples	*n*	χ^2	df	*p*	*r*
All chimps, highest rank	**30**	**20.042**	**2**	**~0**	**0.817**
All chimps, mother's rank	16	1.979	2	0.372	0.3517
All chimps, death rank	**31**	**13.718**	**2**	**0.001**	**0.665**

Bold text indicates a significant relationship. $\alpha = 0.05$

Table 2.18 Results for Kruskal–Wallis tests examining the possible relationship between dominance rank categories and age for adults only

Samples	*n*	χ^2	df	*p*
Adults only, highest rank	15	3.947	2	0.139
Adults only, mother's rank	5	2.00	1	0.157
Adults only, death rank	16	0.122	2	0.122

There are no significant relationships. $\alpha = 0.05$

age versus high rank and death rank are equal to or greater than the effects sizes for dominance rank on skeletal lesion rate. For example, the effect size of skeletal trauma rate versus highest achieved rank in male chimpanzees is $r = 0.645$ (Table 2.15), but this is lower than the calculated effect size of the influence of age on highest achieved dominance rank, which is $r = 0.817$ (Table 2.17). Since age strongly affects most measures of dominance rank in this study, skeletal lesion rate is more likely to be influenced by age than by dominance rank in this sample.

To better control for age, I also examined a sub-sample including only adults. Age does not differ significantly across dominance rank categories in a sub-sample of adults only (Table 2.18). I therefore tested whether skeletal lesion rate differed across dominance rank categories in the adults only sub-sample. I tested for differences in combined-sex as well as single-sex samples, and found no significant differences (Table 2.19).

While the table highlights the limitations of the current study in that sub-sample sizes are often small, results from this study provide additional support for the hypothesis that chimpanzees accumulate skeletal lesions as they age. Continued study of hard tissues from wild primates will improve our understanding of additional factors that are likely to influence the accumulation of skeletal lesions. Only long-term research that integrates multiple lines of evidence can elucidate complex questions about individual selective pressures, life history, and the ways that behaviors affect hard tissues.

2.19 Comparison with Kibale

Comparison of skeletal pathology and trauma between chimpanzee research sites has the potential to inform us about possible differences in sources of selection pressures, morbidity, and mortality. Study of the Gombe skeletal collection as well as

Table 2.19 Results from Kruskal–Wallis tests examining the possible relationship between dominance rank categories and skeletal lesion rates for adult chimpanzees

Samples	n	χ^2	df	p
All adults				
Trauma rate, highest rank	15	4.559	2	0.102
Trauma rate, mother's rank	5	1.184	1	0.277
Trauma rate, death rank	16	4.538	2	0.103
Pathology rate, highest rank	15	0.949	2	0.622
Pathology rate, mother's rank	5	2.105	1	0.147
Pathology rate, death rank	16	3.854	2	0.146
Any lesion rate, highest rank	15	1.85	2	0.397
Any lesion rate, mother's rank	5	2	1	0.157
Any lesion rate, death rank	16	3.499	2	0.174
Adult males				
Trauma rate, highest rank	No comparisons possible; all were high ranking			
Trauma rate, mother's rank	No comparisons possible; all were middle ranking			
Trauma rate, death rank	6	0.484	1	0.487
Pathology rate, highest rank	No comparisons possible; all were high ranking			
Pathology rate, mother's rank	No comparisons possible; all were middle ranking			
Pathology rate, death rank	6	0.455	1	0.5
Any lesion rate, highest rank	No comparisons possible; all were high ranking			
Any lesion rate, mother's rank	No comparisons possible; all were middle ranking			
Any lesion rate, death rank	6	0.784	1	0.376
Adult females				
Trauma rate, highest rank	10	1.021	2	0.6
Trauma rate, mother's rank	3	0.5	1	0.48
Trauma rate, death rank	10	2.813	2	0.245
Pathology rate, highest rank	10	0.173	2	0.917
Pathology rate, mother's rank	3	1.5	1	0.221
Pathology rate, death rank	10	0.926	2	0.629
Any lesion rate, highest rank	10	1.739	2	0.419
Any lesion rate, mother's rank	3	1.5	1	0.221
Any lesion rate, death rank	10	0.114	2	0.945

There are no significant relationships. $\alpha = 0.05$

skeletons from Kibale National Park in Uganda (n = 20 chimpanzees) shows that chimpanzees from both study sites experience high levels of pathology and trauma (Carter et al. 2008). The number of chimpanzees affected by pathology in the Kibale sample also approaches 100% (Ibid.). (A more specific rate describing individuals affected by trauma is not possible to ascertain from the published data as some pathologies are listed categorically and it is unclear whether a single chimpanzee appears in more than one category.) Thirty (30) of 36 (83%) complete skeletons from Gombe exhibit some kind of pathology (bone loss, bone formation, or congenital variation). Eleven of the 12 chimpanzees from Kibale with post-cranial skeletons are affected by trauma (92%) (Ibid.). Of the 36 complete skeletons from Gombe presented here, 25 are affected by trauma (70%).

Some methodological issues worth noting include that the Kibale analysis only reports antemortem trauma, while the Gombe analysis includes both ante- and perimortem traumata. I decided to include both types of trauma in this comparison because their exclusion seems likely to be due to the relatively small sample available from Kibale in which no chimpanzees would be likely to have sustained perimortem trauma (random sampling bias) rather than any real lack of perimortem lesions in chimpanzees from Kibale in general. Both the samples compared in this section are of mixed sex and include multiple age categories. The Gombe sample, being larger, represents a wider range of ages.

Second, I excluded the distal limbs (hands/feet) because of the discrepancy in the ways these elements were counted between the two studies. While Carter and colleagues counted each hand and foot as a unit (scoring each as present if 75% or more of the elements were available for analysis), I counted each individual element (e.g., most animals had two capitates available for analysis). Similar issues surround the ribs and vertebrae, and it seems that no sterna, hyoids, or patellae were available for analysis of traumatic lesions from Kibale as they are not accounted for in the data tables (see below sterno-clavicular joint, however). I comment on trauma to the distal limbs below.

While comparisons between the two study sites are currently limited by small skeletal samples, especially from Kibale (i.e., in the case of several categories of elements, none were affected by traumatic lesions), a few patterns emerge (Table 2.20). The head is the skeletal region most affected by trauma in both skeletal collections (Fig. 2.20), though the rate appears higher at Gombe. Conversely, the hand may be more affected by trauma at Kibale. Approximately 47% of observable hands from Kibale exhibit evidence for skeletal trauma, while 11% of feet were affected. Only 3% of all hand and foot bones from the Gombe sample are affected by trauma. While this may be due in part to differences in scoring methodology, the much higher rate of snare injuries to chimpanzees at Kibale must also play a critical role (Carter et al. 2008).

Table 2.20 Rates of skeletal traumatic lesions at Gombe versus Kibale

	Kibale elements	Number affected	Kibale %	Gombe elements	Number affected	Gombe %
Skull	31	10	32	37	28	77
Clavicle	19	0	0	68	2	3
Scapula	19	0	0	69	5	7
Humerus	22	0	0	69	7	10
Radius	19	2	11	69	2	3
Ulna	19	2	11	70	8	11
Os coxa	19	1	5	67	8	12
Femur	19	0	0	68	7	10
Tibia	17	1	6	67	7	10
Fibula	18	0	0	66	4	6

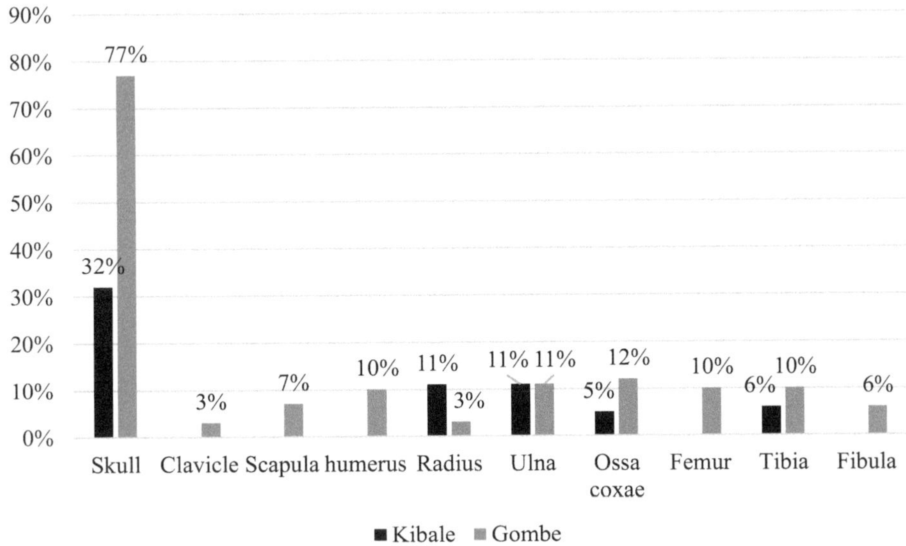

Fig. 2.20 Comparison of skeletal traumata at Gombe versus Kibale

The sample from Kibale analyzed by Carter and colleagues is similar in the rates and types of traumatic lesions relative to an earlier, smaller sample from Gombe (Carter et al. 2008; Jurmain 1989). The same may be said for the rate at which individuals experienced moderate or severe degenerative joint disease unrelated to trauma. While Carter and colleagues highlight low rates for both skeletal samples (Carter et al. 2008), the expanded sample from Gombe offers the chance to expand the analysis (Table 2.21). Fewer than a dozen individuals were previously available for analysis from either Kibale or Gombe, and both samples include regions/joints without any individuals affected by arthropathy (Table 6 from Carter et al. 2008). In the expanded sample from Gombe, however, all of the regions defined by Carter et al. include at least one affected individual. As work at these and other primate research sites continues, we will undoubtedly learn even more about what skeletal lesions can tell us about chimpanzees, particularly across the life span and between demographic categories.

Given comparisons with the expanded sample from Gombe, previously published similarities may be driven largely by sample size. This is not an indictment of previous work, which provided great insight into these animals through the analysis of hard tissues, but support for further long-term research at these and other sites so that our ideas about chimpanzees and other primates, including ourselves, continues to expand.

Table 2.21 Individuals with joints affected by non-traumatic arthropathy

Joint	Kibale individuals	Number affected	Kibale %	Gombe individuals	Number affected	Gombe %
Cervical vertebrae	10	0	0	23	1	4
Thoracic vertebrae	10	0	0	23	2	9
Lumbar vertebrae	10	0	0	23	1	4
TMJ	19	3	16	23	1	4
Sterno-clavicular	10	2	20	23	1	4
Shoulder	11	1	9	23	2	9
Elbow	11	1	9	23	4	17
Hip	10	0	0	23	3	13
Knee	10	1	10	23	3	13
Wrist/hand and ankle/foot	11	0	0	23	4	17

References

Arcadi AC, Wrangham RW (1999) Infanticide in chimpanzees: review of cases and a new within-group observation from the Kanyawara Study Group in Kibale National Park. Primates 40(2): 337–351

Buikstra JE, Ubelaker DH (1994) Standards for data collection from human skeletal remains. Arkansas Archaeological Survey, Fayetteville

Carter M, Pontzer H, Wrangham RW, Peterhans K (2008) Skeletal pathology in *Pan troglodytes schweinfurthii* in Kibale National Park, Uganda. Am J Phys Anthropol 135(4):389–403

Cohen J (1992) A power primer. Psychol Bull 112(1):155–159

Constable JL, Ashley MV, Goodall J, Pusey AE (2001) Noninvasive paternity assignment in Gombe chimpanzees. Mol Ecol 10(5):1279–1300

Cook RD, Weisberg S (1999) Applied regression including computing and graphics. Wiley, New York

Devore J, Peck R (2001) Statistics: the exploration and analysis of data, 4th ed. Duxbury, Pacific Grove

Dixon W (1953) Processing data for outliers. Biometrics 9(1):74–89

Foster MW, Gilby IC, Murray CM, Johnson A, Wroblewski EE, Pusey AE (2009) Alpha male chimpanzee grooming patterns: implications for dominance "style". Am J Primatol 71(2):136–144

Goodall J (1986) The chimpanzees of Gombe: patterns of behavior. Harvard University Press, Cambridge

Hill K, Boesch C, Goodall J, Pusey A, Williams J, Wrangham R (2001) Mortality rates among wild chimpanzees. J Hum Evol 40(5):437–450

IBMCorp (2016) IBM SPSS statistics for windows, 24.0 edn. IBM Corp, Armonk

Jurmain R (1989) Trauma, degenerative disease, and other pathologies among the Gombe chimpanzees. Am J Phys Anthropol 80:229–237

Jurmain R (1997) Skeletal evidence of trauma in African apes, with special reference to the Gombe chimpanzees. Primates 38(1):1–14

Jurmain R, Kilgore L (1998) Sex-related patterns of trauma in humans and great apes. In: Grauer AL, Stuart-MacAdam P (eds) Sex and gender in paleopathological perspective. Cambridge University Press, Cambridge, pp 11–26

Kahlenberg SM, Thompson ME, Muller MN, Wrangham RW (2008) Immigration costs for female chimpanzees and male protection as an immigrant counterstrategy to intrasexual aggression. Anim Behav 76:1497–1509

Kanji GK (1999) 100 statistical tests. SAGE, London

Kirchhoff CA, Wilson ML, Mjungu DC, Raphael J, Kamenya S, Collins DA (2018) Infanticide in chimpanzees: Taphonomic case studies from Gombe. Am J Phys Anthropol 165(1):108–122

Kruskal WH, Wallis WA (1952) Use of ranks in one-criterion variance analysis. J Am Stat Assoc 47(260):583–621

Latimer B (1993) Review of patterns of injury and illness in great apes: a skeletal analysis, by Nancy C. Lovell. J Hum Evol 24(4):335–336

Lovell NC (1990) Patterns of injury and illness in great apes: a skeletal analysis. Smithsonian Institution Press, Washington, DC

Morbeck ME, Galloway A, Sumner DR (2002) Getting old at Gombe: skeletal aging in wild-ranging chimpanzees. In: Erwin JM, Hof PR (eds) Aging in non-human primates. Karger, Basel, pp 48–62

Muller MN, Wrangham RW (2004) Dominance, cortisol and stress in wild chimpanzees (*Pan troglodytes schweinfurthii*). Behav Ecol Sociobiol 55:332–340

Murray CM, Eberly LE, Pusey AE (2006) Foraging strategies as a function of season and rank among wild female chimpanzees (*Pan troglodytes*). Behav Ecol 17(6):1020–1028

Novak SA, Hatch MA (2009) Intimate wounds: craniofacial trauma in women and female chimpanzees. In: Muller MN, Wrangham RW (eds) Sexual coercion in primates and humans: an evolutionary perspective on male aggression against females. Harvard University Press, Cambridge, pp 322–345

Pusey AE, Oehlert GW, Williams JM, Goodall J (2005) Influence of ecological and social factors on body mass of wild chimpanzees. Int J Primatol 26(1):3–31

Pusey AE, Williams J, Goodall J (1997) The influence of dominance rank on reproductive success of female chimpanzees. Science 277:828–831

Rando C, Waldron T (2012) TMJ osteoarthritis: a new approach to diagnosis. Am J Phys Anthropol 148:45–53

Rosenthal R, DiMatteo MR (2001) Meta-analysis: recent developments in quantitative methods for literature reviews. Annu Rev Psychol 52(1):59–82

Thompson ME, Muller MN, Wrangham RW, Lwanga JS, Potts KB (2009) Urinary C-peptide tracks seasonal and individual variation in energy balance in wild chimpanzees. Horm Behav 55:299–305

Williams JM, Lonsdorf EV, Wilson ML, Schumacher-Stankey J, Goodall J, Pusey AE (2008) Causes of death in the Kasekela chimpanzees of Gombe National Park, Tanzania. Am J Primatol 70:766–777

Wrangham RW (2000) Why are male chimpanzees more gregarious than mothers? A scramble competition hypothesis. In: Primate males: causes and consequences of variation in group composition. Cambridge University Press, Cambridge, pp 248–258

Wroblewski EE, Murray CM, Keele BF, Schumacher-Stankey JC, Hahn BH, Pusey AE (2009) Male dominance rank and reproductive success in chimpanzees, *Pan troglodytes schweinfurthii*. Anim Behav 77:873–885

Chapter 3
Discussion

3.1 Sex Differences in Skeletal Lesion Rate

Previous work on sex differences in skeletal lesion rates primes us to ask questions about how differing selective pressures on females versus males might affect the hard tissues. For example, increased prevalence of dental caries among human females relative to males is common and widespread in the archaeological record (Lukacs and Largaespada 2006; Lukacs and Thompson 2008). Studies of human biology also highlight differences in immune reactivity, with females characterized as having stronger and more effective immune responses to infectious diseases, an adaptation that may also be relevant for other ape species (Ortner 1998). While many studies that specifically examine sex differences in skeletal lesions report few or no significant differences between the sexes (Vanna 2007; Delgado-Darias et al. 2005; Pechenkina et al. 2002; Roberts et al. 1998; Storey 1998; Larsen 1998), some contexts provide evidence for sex differences in skeletal lesions that elucidate possible variation in selective pressures.

Jurmain's (1997) examination of skeletal trauma in the African apes suggests that falls are a significant source of morbidity, and he noted that the forelimb seemed especially vulnerable to injury. He also noted that high rates of inter-individual aggression could be a possible source of skeletal injury, particularly for males. His study provided no evidence, however, for significant sex differences in post-cranial antemortem skeletal trauma (Jurmain 1997; Jurmain and Kilgore 1998). In contrast, cranial trauma exhibited distinct sex differences, with males exhibiting higher rates of cranial trauma (Jurmain and Kilgore 1998), and lesion placement differing between males and females. While females exhibit more traumata to the cranial vault, males are more likely to sustain facial fractures (Jurmain and Kilgore 1998; Novak and Hatch 2009). The same pattern of cranial lesion distribution is true for the Gombe chimpanzee skeletons.

C. A. Kirchhoff, *Life and Death in the Gombe Chimpanzees*,
Developments in Primatology: Progress and Prospects,
https://doi.org/10.1007/978-3-030-18355-4_3

In addition to fractures, bite wounds also exhibit a possible difference in lesion distribution. While Jurmain found that only male chimpanzees exhibited skeletal evidence for bite wounds (Jurmain and Kilgore 1998), the report is ambiguous regarding which skeletal elements were evaluated for this presence of these lesions. In contrast, both males and females in the Gombe chimpanzee skeletal sample exhibit bite wounds. While there are only five adult animals thus affected, the possible sex differences in the location of these lesions would be worth continued investigation. There are three males who likely sustained facial bite wounds (Charlie, MacDee, Vincent), and one male who likely sustained bite wounds to the hindlimb (Vincent, again). There are two females who likely sustained bite wounds (Madam Bee, Patti), and both cases are injuries to the post-cranial skeleton.

An important distinction between this study and previous, valuable work by Jurmain is the nature of the sample. Jurmain's large studies include data from museum collections, which consist primarily in wild-shot animals (Jurmain and Kilgore 1998; Harman 2005) who all have the same cause of death. The Gombe skeletal collection, however, is comprised of animals who experienced a wide range of causes of death, with illness and conspecific aggression (rather than human intervention) being some of the most common causes of death (Williams et al. 2008). A wild-shot population cannot include perimortem conspecific-inflicted trauma, but the Gombe chimpanzee skeletal collection does. This is one example of the ways in which populations of living animals versus natural death samples versus museum skeletal samples of wild primates may differ (Buikstra 1975).

Differences between sample types may help explain why female chimpanzees from Gombe exhibit a higher rate of skeletal trauma to the trunk compared to males. Trunk injuries documented in this collection may all be attributed to fractures rather than to punctures/bite wounds. Fractures sustained as a result of falls versus agonistic encounters are difficult to distinguish without direct evidence linking a particular skeletal lesion to an observed incident. While there is evidence that adult female chimpanzees are more arboreal than males (Doran 1993), and might therefore be at increased risk for falls, skeletal trauma rates in the Gombe chimpanzee skeletal collection are also likely influenced by the female infanticide victims, who exhibit very high rates of skeletal trauma—including to the trunk.

In Chap. 2, I described the following statistical differences in pathological lesion rates: females exhibit more forelimb pathologic lesions while males exhibit more pathologic lesions of the trunk and hindlimb. There were no other statistical differences in the pathologic lesion rate in other body regions. Few previous studies contrast skeletal pathology rates in female versus male nonhuman primates (see, e.g., Lovell 1991; Schultz 1956, which do not examine sex differences). A broader interpretive framework is necessary to understand the possible biological significance of the statistical differences between skeletal lesion rates documented in the Gombe chimpanzee skeletal collection. Continuing to investigate possible differences between sexes and age categories in skeletal lesion rate and placement may allow further exploration of the ways selective pressures differentially impact females versus males of diverse ages.

3.2 The Complex Relationship Between Dominance Rank and Skeletal Lesions

Dominance rank mediates chimpanzees' social interactions and is an important source of variation for many life history traits and measures of health. Previous work suggests that the relationship between stress and dominance rank in primates is complex (Honess and Marin 2006), though studies of cortisol levels indicate that that under certain circumstances stress levels vary predictably for individuals of different ranks (see Chap. 1). I therefore tested whether another measure would match previously described variation in stress levels across dominance rank categories. Towards this end, I analyzed skeletal lesions and how they might vary according to dominance rank.

Some of the ways dominance rank affects chimpanzees include that high-ranking females have improved access to feeding resources compared to their lower-ranking counterparts (Pusey et al. 1997; Thompson et al. 2007; Murray et al. 2006). In addition, high-ranking females are more likely to have ranges near the center of their group's territory, making them less vulnerable to inter-group aggression (Nishida 1989; Williams et al. 2004). High-ranking females also achieve larger body masses (Pusey et al. 2005) and higher rates of reproductive success compared to lower-ranking females (Pusey et al. 1997). While body mass doesn't vary between ranks for males (Pusey et al. 2005), high-ranking males have preferential access to high-quality foraging sites relative to lower-ranking chimpanzees (Wrangham 1975). Like females, high-ranking males also achieve higher rates of reproductive success (Wroblewski et al. 2009).

Previous examinations of the role dominance rank plays in the lives of chimpanzees suggest that rank is likely to have a strong influence on other variables that measure health and stress. In wild primates, levels of the hormone cortisol can serve as a proxy for stress levels, and cortisol levels do vary between chimpanzees of different ranks. High-ranking males experience elevated stress levels, but they are not always the most stressed. While high-ranking males have elevated cortisol levels compared to conspecifics (Muller and Wrangham 2004) and devote a significant amount of time to maintaining their rank (and therefore less time feeding) (Thompson et al. 2009), immigrant females also have elevated cortisol levels that likely result from the conspecific aggression with which they must contend while integrating into a new community (Kahlenberg et al. 2008a).

This aligns with the framework, developed by Sapolsky (2005), for predicting which primates in a group experience the highest levels of stress. The predictions incorporate variation in social characteristics that influence the effect of being high versus low ranking on an individual animal. Sapolsky hypothesizes that high-ranking animals experience the highest levels of stress in populations where their rank must be maintained through frequent physical agonistic encounters, dominant animals perceive most interactions as potential threats, a cooperative breeding system is used, and/or when ranks are unstable. In contrast, low-ranking animals experience the highest levels of stress in populations where the hierarchy is maintained through frequent intimidation, and subordinates have little access to coping mechanisms such as affiliative behaviors. Low-ranking animals are also the most stressed in groups with stable hierarchies, groups where it is difficult to avoid the dominant animals, and/or there are few alternatives to direct competition (Ibid.).

This framework provides insight into why both high-ranking males and low-ranking, immigrant females experience elevated cortisol levels compared to other chimpanzees. High-ranking male chimpanzees often maintain their dominant position through physical agonistic encounters (Thompson et al. 2009; Muller and Mitani 2005; Goodall 1986), and ranks are subject to change throughout the course of an animal's life, particularly for males (Goodall 1986; Foerster et al. 2016). Low-ranking chimpanzees do have recourse to coping mechanisms and the ability to avoid dominant animals (e.g., by "fissioning" to forage alone) (Wrangham 2000; Murray et al. 2006). These factors predict that high-ranking chimpanzees, especially males, will be the most stressed animals in the group. A more complex dynamic is suggested, however, by variation in dominance styles or personalities (Foster et al. 2009; Honess and Marin 2006). A high-ranking male with a "despotic" style (*sensu* Sapolsky 2005) of dominating his conspecifics may, for example, experience higher levels of stress compared to a dominant male who maintains his rank through the use of grooming and other affiliative behaviors (Foster et al. 2009). Intimidation is also a tactic used by high-ranking males to dominate subordinates (Ibid.), and females attempting to join a new group may opt not to avoid high-ranking conspecifics (Kahlenberg et al. 2008a; Williams et al. 2002b). For males, high rank comes with an energetic cost (Thompson et al. 2009), which also contributes to the ways in which relative stress levels between chimpanzees of different rank categories are complex.

While Sapolsky (2005) predicts that the degree and style of despotism can have a greater impact on higher ranking males when frequent physical assertion of dominance takes place, lower-ranking males are likely to be more stressed if the hierarchy is maintained through frequent intimidation. Depending on the "personalities" or "leadership styles" of dominant male chimpanzees (Foster et al. 2009), the dominance style (*sensu* Sapolsky 2005) of a given chimpanzee community may actually vary over the course of a chimpanzee's life. Relative levels of stress between higher-versus lower-ranking chimpanzees might therefore also vary.

In addition, when dominant animals are highly affiliative, Sapolsky (2005) predicts that lower-ranking animals will experience more stress. This can, however, be mitigated in circumstances that allow for subordinates to partake in "coping and support" behaviors (e.g., grooming, reconciliation) (Ibid.), which decrease stress levels for lower-ranking animals (Honess and Marin 2006; Sapolsky 2005). Social bonds between male chimpanzees are often reinforced through these types of affiliative behaviors, likely reducing the level of stress experienced by subordinates (De Waal and van Roosmalen 1979; Silk 2002; Kutsukake and Castles 2004).

Low-ranking and new immigrant females have elevated cortisol levels relative to high-ranking conspecifics (Thompson et al. 2010). Females joining a new group experience aggression from resident females that is likely related to food competition (Kahlenberg et al. 2008a, b), and often attempt to socialize with high-ranking males in order to establish group membership (Kahlenberg et al. 2008a). This potentially exposes the female to increased agonistic encounters, both from competing females and from coercive males, or males competing vigorously for mating opportunities with her (Ibid.), in contrast to females foraging alone (or with dependent offspring) (Pusey et al. 1997) who may avoid aggressive males. Female chimpanzees at Gombe may spend more time alone than females from other regions (Williams et al. 2002a). Additional possible stressors for low-ranking females include making use of lower-quality foraging ranges that result in the need to spend more time acquiring food and eating less desirable foods (Murray et al. 2006, 2007). Lower-ranking females also more frequently make use of the periphery of their group's territory (Murray et al. 2007; Thompson et al. 2007), which may make them more vulnerable to attacks from a neighboring group (Williams et al. 2004).

I predicted that the higher rates of stress experienced by high-ranking male chimpanzees would be reflected by a higher incidence of skeletal lesions compared to conspecifics. Similarly, I hypothesized that low-ranking females would also exhibit higher rates of lesions. While my predictions were informed by studies on cortisol levels (discussed above), my findings highlight the complex interactions between rank and stress.

I found some evidence that the rate of skeletal lesions in high-ranking males is higher compared to other demographic groups, but the rate of skeletal lesions in the Gombe chimpanzees is more closely associated with age. Since dominance rank is such an important variable for chimpanzees, it is worthwhile to examine possible reasons why the link between dominance rank and skeletal lesions is weak.

First, primates are characterized by complex learned behaviors and social interactions as well as behavioral plasticity (reviewed as related to stress and dominance rank in Honess and Marin 2006). This is particularly true of the great apes, and therefore relationships between factors that are themselves complex, such as dominance rank and stress, will necessarily be multifaceted. For chimpanzees, this is already clear from hormonal studies: while high-ranking males exhibit higher cortisol levels, they do not always have the highest cortisol levels (Kahlenberg et al. 2008a; Muller and Wrangham 2004).

In addition, multiple factors influence stress levels for chimpanzees, which may not be accurately represented by or have a simple relationship with skeletal lesions. For example, stress in female chimpanzees is influenced by life history stage as well as conspecific aggression. Lactating females experience higher stress levels when fruit is scarce, and estrous females experience higher stress levels due to coercive behavior from males. Similarly, agonistic behavior (which may result in skeletal lesions) is stressful for both the aggressor and the target (Thompson et al. 2010). Social instability can result in increased stress for chimpanzees (Honess and Marin 2006; Sapolsky 2005), particularly for males (Sapolsky 2005) who compete directly for social status (Foerster et al. 2016; Muller and Mitani 2005; Thompson et al. 2009). This type of stressor, as well as other, long-term influences, may result in chronic stress, which is known to result in decreased cortisol production (Honess and Marin 2006). Clearly, the relationship between stress and rank is multifactorial in chimpanzees.

This study is also limited by sample size. While it is the largest collection of wild chimpanzee skeletons with life history data available, the sample presented here includes fewer than 50 animals. This makes it difficult to control for age, which influences rank (e.g., all sub-adults are low ranking, and older males tend to decrease in rank as they age (Goodall 1986; GSRC)). The sample is also limited in that not all age categories (as defined by Goodall (1986)) are included in this sample, nor is the skeletal sample entirely representative of either the population of living chimpanzees at Gombe (Gombe Stream Research Center Annual Report 2016, unpublished data) or the analyzed sample of chimpanzees from Gombe who have died (described by Williams and colleagues in 2008, see Chap. 2). The ways in which the skeletal sample differs demographically from the other samples include differing sex ratios and unequal distribution of age categories. While these discrepancies may have the potential to improve our understanding of the ways in which skeletal collections are likely to differ from living populations, they may also contribute to obscuring possible relationships between skeletal lesions and dominance rank. Being able to control for age in a future, larger sample where refined age categories might be considered separately (Wright and Yoder 2003) may allow us to examine the effects of dominance rank on skeletal lesions more closely.

The "osteological paradox" posits that multiple interpretations of skeletal lesions may be equally valid, in the absence of supporting evidence that would favor a particular explanation (Wood et al. 1992). For example, the osteological paradox challenges conventional interpretations of skeletal lesions which begin with the premise that individuals with the most or the most severe lesions were necessarily the least healthy and/or most stressed within a population. It may, in some cases, be more parsimonious to assume that such individuals represent the heartiest individuals within their group. The ability to survive with a chronic illness long enough for it to remodel the skeleton may be evidence for an increased ability to buffer against stressors such as illness, injury, or malnutrition. Skeletons with fewer and/or less severe lesions may therefore represent individuals who were less well equipped to mitigate the effects of stress and died due to an acute illness or injury before it left any skeletal signs. In addition, heterogeneity in individual responses to stress, disease, or other selective factors means that the relationship between skeletal lesions and overall health may not always be direct (DeWitte and Stojanowski 2015; Wood et al. 1992).

This may relate to the evidence presented here that skeletal lesion rate increases with age: chimpanzees best equipped to survive illness and injury may live longer (and accumulate more skeletal lesions) (see also Jurmain 1997, 1989). Rank and adult age do not seem to be related in this sample, so the Gombe chimpanzee skeletons do not provide evidence that high rank contributes to longevity in the skeletal sample. Given the restrictions of sample size (n = 15 adult animals with well-documented highest achieved rank), however, further work on a future, larger sample might prove fruitful, particularly when analyzing males and females separately. Behavioral research provides evidence that high-ranking females attain longer lifespans than lower-ranking conspecifics (Pusey et al. 1997).

A larger sample would also provide more opportunity to examine dominance rank in greater depth. For example, male chimpanzees are commonly assigned ordinal ranks, and this may prove to be a better measure of rank than the three categories employed for this analysis. In addition, more data on mother's rank at the time of an individual's birth/infancy may provide further insight into the relationship between stress, health, rank, and skeletal lesions. This sample includes only six adult chimpanzees whose mother's rank at (or near) their time of birth is known. Previous work has shown that mother's high rank can have a significant influence on her offspring, including earlier attainment of sexual maturity, improved nutrition resulting in larger body sizes, and better offspring survival (Pusey et al. 1997; Foerster et al. 2016).

As the Gombe chimpanzees continue to be observed, more animals will have family tree information available, including paternity data (Wroblewski et al. 2009). Analyses of the potential effects of paternal rank as well as maternal rank may be of interest. The ways in which kinship and support networks influence stress and skeletal lesions are also worth examining, particularly given that these are prime coping mechanisms to mitigate the effects of stress in subordinate animals (Sapolsky 2005).

This sample, while restricted in terms of sample size, did reveal trends not evident in an earlier sub-set of the sample (animals who died before 1986). It is reasonable to predict that a future, larger sample would demonstrate additional trends and relationships that are not currently detectable. Future work on even longer-term trends will provide additional insights into this long-lived species.

For example, many of the animals in the current skeletal sample have estimated ages. These estimates are considered robust as they are based on decades of observation of living animals, but variation in skeletal lesions and tooth wear (see, e.g., the variation in tooth wear and antemortem tooth loss in Patti versus Melissa and Miff or Goblin versus Hugo described in Chap. 1) suggest that re-visiting these estimates when a larger skeletal sample of animals with known ages is available would be worthwhile. Such an endeavor would be of particular interest given the strong relationship between age and the accumulation of skeletal lesions described here, in order to provide additional insight into this correlation, as well as possible sources of variation between animals of similar age.

Future lines of inquiry for further exploring possible connections between dominance rank and skeletal data include the analysis of other markers of stress or developmental disturbance such as fluctuating asymmetry (Palmer 1994; Palmer and Strobeck 1986; Romero 2018; Romero and Terhune 2018) and enamel hypoplasia (Guatelli-Steinberg and Lukacs 1999; Goodman and Rose 1990; Skinner and Hopwood 2004). Both fluctuating asymmetry and enamel defects are used to measure developmental instability, or stress during the period of growth and development. Linking these measures with dominance rank data, as well as data on other skeletal and dental lesions, would provide additional insight into which chimpanzees might be best equipped to achieve high rank (and its benefits, such as increased reproductive success). Are chimpanzees who experienced low levels of developmental instability more likely to achieve high rank? Or, are chimpanzees with higher levels of, e.g., fluctuating asymmetry, demonstrating increased resilience or some other quality that is correlated with high rank?

Data from the Gombe chimpanzee skeletal collection are beginning to be integrated with data from behavioral observation, soft tissue lesions (Terio et al. 2011), hormone analysis, and other lines of evidence. Analyzing these disparate sources of information in a holistic way is the best way to further our understanding of chimpanzees in particular, primates in general, and, by extension, ourselves. Continued integration of data from the hard tissues with other kinds of analyses extends our ability to tell life stories from bones.

References

Buikstra JE (1975) Healed fratures in *Macaca mulatta*: Age, sex, and symmetry. Folia Primatol 23:140–148

De Waal FB, van Roosmalen A (1979) Reconciliation and consolation among chimpanzees. Behav Ecol Sociobiol 5(1):55–66

Delgado-Darias T, Velasc-Vázquez J, Arnay-de-la-Rosa M, Martín-Rodríguez E, González-Reimers E (2005) Dental caries among the prehispanic population from gran Canaria. Am J Phys Anthropol 128(3):560–568

DeWitte SN, Stojanowski CM (2015) The osteological paradox 20 years later: Past perspectives, future directions. J Archaeol Res 23(4):397–450

Doran DM (1993) Sex differences in adult chimpanzee positional behavior: The influence of body size on locomotion and posture. Am J Phys Anthropol 91(1):99–115

Foerster S, Franz M, Murray CM, Gilby IC, Feldblum JT, Walker KK, Pusey AE (2016) Chimpanzee females queue but males compete for social status. Sci Rep 6:35404

Foster MW, Gilby IC, Murray CM, Johnson A, Wroblewski EE, Pusey AE (2009) Alpha male chimpanzee grooming patterns:Iimplications for dominance “style”. Am J Primatol 71(2):136–144

Goodall J (1986) The Chimpanzees of Gombe: Patterns of Behavior. Harvard University Press, Cambridge

Goodman AH, Rose JC (1990) Assessment of systemic physiological perturbations from dental enamel hypoplasias and associated histological structures. Am J Phys Anthropol 33:59–110

Guatelli-Steinberg D, Lukacs JR (1999) Interpreting sex differences in enamel hypoplasia in human and non-human primates: Developmental, environmental, and cultural considerations. Yearb Phys Anthropol 42:73–126

Harman MP (2005) Primate Collection in the Powell-Cotton Museum. Powell-Cotton Museum, Birchington

Honess PE, Marin CM (2006) Behavioural and physiological aspects of stress and aggression in nonhuman primates. Neurosci Biobehav Rev 30(3):390–412

Jurmain R (1989) Trauma, degenerative disease, and other pathologies among the Gombe chimpanzees. Am J Phys Anthropol 80:229–237

Jurmain R (1997) Skeletal evidence of trauma in African apes, with special reference to the Gombe chimpanzees. Primates 38(1):1–14

Jurmain R, Kilgore L (1998) Sex-related patterns of trauma in humans and great apes. In: Grauer AL, Stuart-MacAdam P (eds) Sex and Gender in Paleopathological Perspective. Cambridge University Press, Cambridge, pp 11–26

Kahlenberg SM, Thompson ME, Muller MN, Wrangham RW (2008a) Immigration costs for female chimpanzees and male protection as an immigrant counterstrategy to intrasexual aggression. Anim Behav 76:1497–1509

Kahlenberg SM, Thompson ME, Wrangham RW (2008b) Female competition over core areas in *Pan troglodytes schweinfurthii*, Kibale National Park, Uganda. Int J Primatol 29(4):931

Kutsukake N, Castles DL (2004) Reconciliation and post-conflict third-party affiliation among wild chimpanzees in the Mahale Mountains, Tanzania. Primates 45(3):157–165

Larsen CS (1998) Gender, health, and activity in foragers and farmers in the American southeast: implications for social organization in the Georgia Bight. In: Grauer AL, Stuart-Macadam P (eds) Sex and Gender in Paleopathological Perspective. Cambridge University Press, Cambridge, pp 165–187

Lovell NC (1991) An evolutionary framework for assessing illness and injury in nonhuman primates. Yearb Phys Anthropol 34:117–155

Lukacs JR, Largaespada LL (2006) Explaining sex differences in dental caries prevalence: saliva, hormones, and "life-history" etiologies. Am J Hum Biol 18(4):540–555

Lukacs JR, Thompson LM (2008) Dental caries prevalence by sex in prehistory: magnitude and meaning. In: Irish JD, Nelson GC (eds) Technique and Application in Dental Anthropology. Cambridge University Press, Cambridge, pp 136–177

Muller MN, Mitani JC (2005) Conflict and cooperation in wild chimpanzees. In: Advances in the Study of Behavior, vol 35. Academic Press, New York, pp 275–331

Muller MN, Wrangham RW (2004) Dominance, cortisol and stress in wild chimpanzees (*Pan troglodytes schweinfurthii*). Behav Ecol Sociobiol 55:332–340

Murray CM, Eberly LE, Pusey AE (2006) Foraging strategies as a function of season and rank among wild female chimpanzees (*Pan troglodytes*). Behav Ecol 17(6):1020–1028

Murray CM, Mane SV, Pusey AE (2007) Dominance rank influences female space use in wild chimpanzees, *Pan troglodytes*: towards an ideal despotic distribution. Anim Behav 74(6):1795–1804

Nishida T (1989) Social interaction between resident and immigrant female chimpanzees. In: Heltne P, Marquardt L (eds) Understanding Chimpanzees. Harvard University Press, Cambridge, pp 68–89

Novak SA, Hatch MA (2009) Intimate wounds: Craniofacial trauma in women and female chimpanzees. In: Muller MN, Wrangham RW (eds) Sexual Coercion in Primates and Humans: An Evolutionary Perspective on Male Aggression Against Females. Harvard University Press, Cambridge, pp 322–345

Ortner DJ (1998) Male-female immune reactivity and its implications for interpreting evidence in human skeletal paleopathology. In: Grauer AL, Stuart-Macadam P (eds) Sex and Gender in Paleopathological Perspective. Cambridge University Press, Cambridge, pp 79–92

Palmer A (1994) Fluctuating asymmetry analyses: A primer. In: Markow T (ed) Developmental Instability: Its Origins and Evolutionary Implications. Kluwer, Dordecth, pp 335–364

Palmer AR, Strobeck C (1986) Fluctuating asymmetry: Measurement, analysis, patterns. Annu Rev Ecol Syst 17(1):391–421

Pechenkina EA, Benfer RA Jr, Zhijun W (2002) Diet and health changes at the end of the Chinese Neolithic: The Yangshao/Longshan transition in Shaanxi province. Am J Phys Anthropol 117(1):15–36

Pusey AE, Williams J, Goodall J (1997) The influence of dominance rank on reproductive success of female chimpanzees. Science 277:828–831

Pusey AE, Oehlert GW, Williams JM, Goodall J (2005) Influence of ecological and social factors on body mass of wild chimpanzees. Int J Primatol 26(1):3–31

Roberts CA, Lewis ME, Boocock P (1998) Infectious disease, sex, and gender: the complexity of it all. In: Grauer AL, Stuart-Macadam P (eds) Sex and Gender in Paleopathological Perspective. Cambridge University Press, Cambridge, pp 93–113

Romero AN (2018) A comparison of craniofacial asymmetry in *Gorilla gorilla gorilla* and *Pan troglodytes troglodytes*. M.A. Thesis, University of Arkansas, Fayetteville

Romero AN, Terhune CE (2018) A comparison of fluctuating asymmetry models in non-human primate crania. Am J Phys Anthropol 165(S66):230

Sapolsky RM (2005) The influence of social hierarchy on primate health. Science 308(5722):648–652

Schultz AH (1956) The occurrence and frequency of pathological and teratological conditions and of twinning among non-human primates. Primatologia 1:965–1014
Silk JB (2002) The form and function of reconciliation in primates. Annu Rev Anthropol 31(1):21–44
Skinner MF, Hopwood D (2004) Hypothesis for the causes and periodicity of repetitive linear enamel hypoplasia in large, wild African (*Pan troglodytes* and *Gorilla gorilla*) and Asian (*Pongo pygmaeus*) apes. Am J Phys Anthropol 123:216–235
Storey R (1998) The mothers and daughters of a patrilineal civilization: The health of females among the late classic Maya of Copán, Honduras. In: Grauer AL, Stuart-Macadam P (eds) Sex and Gender in Paleopathological Perspective. Cambridge University Press, Cambridge, pp 133–148
Terio KA, Kinsel MJ, Raphael J, Mlengeya T, Lipende I, Kirchhoff CA, Gilagiza B, Wilson ML, Kamenya S, Estes JD, Keele BF, Rudicell RS, Liu W, Patton S, Collins A, Hahn BH, Travis DA, Lonsdorf EV (2011) Pathologic lesions in chimpanzees (*Pan trogylodytes schweinfurthii*) from Gombe National Park, Tanzania, 2004–2010. J Zoo Wildl Med 42(4):597–607
Thompson ME, Kahlenberg SM, Gilby IC, Wrangham RW (2007) Core area quality is associated with variance in reproductive success among female chimpanzees at Kibale National Park. Anim Behav 73(3):501–512
Thompson ME, Muller MN, Wrangham RW, Lwanga JS, Potts KB (2009) Urinary C-peptide tracks seasonal and individual variation in energy balance in wild chimpanzees. Horm Behav 55:299–305
Thompson ME, Muller MN, Kahlenberg SM, Wrangham RW (2010) Dynamics of social and energetic stress in wild female chimpanzees. Horm Behav 58(3):440–449
Vanna V (2007) Sex and gender related health status differences in ancient and contemporary skeletal populations. Pap Inst Archacol 18:114–147
Williams JM, Liu H-Y, Pusey AE (2002a) Costs and benefits of grouping for female chimpanzees at Gombe. In: Boesch C, Hohmann G, Marchant LF (eds) Behavioural Diversity in Chimpanzees and Bonobos. Cambrige University Press, Cambridge, pp 192–203
Williams JM, Pusey AE, Carlis JV, Farm BP, Goodall J (2002b) Female competition and male territorial behaviour influence female chimpanzees' ranging patterns. Anim Behav 63(2):347–360
Williams JM, Oehlert GW, Carlis JV, Pusey AE (2004) Why do male chimpanzees defend a group range? Anim Behav 68(3):523–532
Williams JM, Lonsdorf EV, Wilson ML, Schumacher-Stankey J, Goodall J, Pusey AE (2008) Causes of death in the Kasekela chimpanzees of Gombe National Park, Tanzania. Am J Primatol 70:766–777
Wood JW, Milner GR, Harpending HC, Weiss KM, Cohen MN, Eisenberg LE, Hutchinson DL, Jankauskas R, Česnys G (1992) The osteological paradox: problems of inferring prehistoric health from skeletal samples [and comments and reply]. Curr Anthropol 33(4):343–370
Wrangham RW (1975) The Behavioral Ecology of Chimpanzees in the Gombe Sream National Park. Ph.D. Thesis, University of Cambridge, Cambridge
Wrangham RW (2000) Why are male chimpanzees more gregarious than mothers? A scramble competition hypothesis. In: Primate Males: Causes and Consequences of Variation in Group Composition. Cambridge University Press, Cambridge, pp 248–258
Wright LE, Yoder CJ (2003) Recent progress in bioarchaeology: Approaches to the osteological paradox. J Archaeol Res 11(1):43–70
Wroblewski EE, Murray CM, Keele BF, Schumacher-Stankey JC, Hahn BH, Pusey AE (2009) Male dominance rank and reproductive success in chimpanzees, *Pan troglodytes schweinfurthii*. Anim Behav 77:873–885

Index

C. A. Kirchhoff, *Life and Death in the Gombe Chimpanzees*, Developments in Primatology: Progress and Prospects,
https://doi.org/10.1007/978-3-030-18355-4

FSC
www.fsc.org
MIX
Papier | Fördert
gute Waldnutzung
FSC® C083411

Kolibri 360 GmbH
Ferdinand-Jühlke-Straße 7
99095 Erfurt, Deutschland
produktsicherheit@kolibri360.de